AF540141

Chemistry of Transition Elements

Chemistry of Transition Elements

M. SATAKE
Faculty of Engineering
Fukui University
Fukui 910, Japan

Y. MIDO
Dept. of Chemistry
Kobe University
Kobe 657, Japan

Consulting Editors

S.A. IQBAL
Dept. of Chemistry
Saifia P.G. College of
Science and Education
Bhopal, INDIA

M.S. SETHI
A.R.S.D. College
University of Delhi
New Delhi-INDIA

Discovery Publishing House
New Delhi (INDIA)

Reprinted - 2019

First Published - 1994

ISBN: 978-81-7141-243-3

© Author

Chemistry of Transition Elements

Published by:

DISCOVERY PUBLISHING HOUSE PVT. LTD.
4383/4B, Ansari Road, Darya Ganj
New Delhi-110 002 (India)
Phone: +91-11-23279245, 23253475; 43596065
E-mail: discoverybooksindia@gmail.com
discoverypublishinghouse@gmail.com
web: www.discoverypublishinggroup.com

Printed at:
Infinity Imaging Systems
Delhi

Preface

The teaching of Chemistry at the introductory stage becomes each day a more challenging task as the subject matter becomes more diverse and more complex. These challenges have evoked a series of responses – the present set of introductory chemistry monographs is one such. The teaching of chemistry recognises a number of problems that confront those who select text books. In order to overcome these problems, this volume "**Chemistry of Transition Elements**" – one of about fifty in the Chemistry Monograph Series – is introduced. Each volume is independent of the others deals with one of Chemistry topics and constitutes a complete entity. Each volume is more comprehensive than can be possible in a single volume text. It is intended to provide a range of topics to cover most undergraduate and chemistry main courses of study. These volumes can be used to enrich the more conventional courses of study.

Suggestions for improvement are welcome and shall be gratefully acknowledged.

Authors

Contents

Contents

CHAPTER 1

The Properties of Transition Elements

DEFINITION

A transition element may be defined as one which possesses partially filled *d*-orbitals in its penultimate shell. This conceptual definition is useful as it enables us to recognize a transition element merely by looking at its electronic configuration. This definition excludes zinc, cadmium and mercury from the transition elements as they do not have a partially filled *d*-orbital. However, they are also considered as transition elements, because their properties are an extension of the properties of transition elements. In fact, the zinc group serves as a bridge between the transition elements and the representative elements. There transition series are known; they are :

FIRST SERIES	(3*d* series)									
Element	Sc	Ti	V	Cr	Mn	Fe	Co	Ni	Cu	Zn
Atomic No.	21	22	23	24	25	26	27	28	29	30

SECOND SERIES	(4*d* series)									
Element	Y	Zr	Nb	Mo	Tc	Ru	Rh	Pd	Ag	Cd
Atomic No.	39	40	41	42	43	44	45	46	47	48

THIRD SERIES	(5*d* series)									
Element	La	Hf	Ta	W	Re	Os	Ir	Pt	Au	Hg
Atomic No.	57	72	73	74	75	76	77	78	79	80

PROPERTIES

The presence of incomplete inner orbitals results in certain characteristic amongst the transition elements. Some of the general characteristics are listed below :

1. The transition elements show some well-defined horizontal similarities in physical and chemical properties, in contrast the *s*-and *p*-block elements, as well as typical vertical group relationship.
2. Their hardness along with high melting and boiling points suggest strong bonding. The metals possess high density and have close packed structures. They are malleable and ductile, and conduct heat and electricity.
3. There is very little difference in energy between the $(n\text{-}1)d$ and the *ns* electrons, and electrons from both levels can be used in bonding. As a result, the transition elements exhibit variable oxidation states in their compounds. However, there are some exceptions (Sc, Y, Zn, Cd, etc.).
4. They are all metals, varying from moderately electropositive to noble. They form alloys with one another.
5. Transition metal ions generally possess unpaired electrons and are coloured and paramagnetic.
6. Complex ion formation is a typical property of transition elements. All the conditions for the formation of complex ions, viz. small size, high charge and availability of vacant orbitals, are fulfilled by transition elements.

   Not only are the transition metal ions, relatively smaller than the non-transition metal ions, but due to the poor shielding effect of *d*-electrons, the nuclear charge attracting the ligands is greater than expected. Thus, they attract ligands more than the non-transition metal ions of the same size and charge and, therefore, from more stable complexes.
7. The transition elements form compounds with Lewis acid ligands like CO, NO, etc. In these compounds, the elements exist in low, zero or even formally negative oxidation states.
8. The transition elements exhibit catalytic activity. They are wonderful catalysts for hydrogenation (Ni, Pd), oxidation (Pt, V_2O_5), pyrolysis, dehydration, etc.
9. They from *interstitial compounds* with non-metals like C, B, N, H, etc. Properties of the metal are often considerably

altered. The interstial compounds are often hard, have metallic appearance, high melting and boiling points, are good conductors, and are generally chemically inert except to oxidizing agents. Their structure depends upon the non-metallic atom and its amount present.

10. Due to the comparable stability of the metal ions in different oxidation states, the transition elements form *non-stoichiometric compounds.* In several cases there is a gradual change in the properties of the compounds from one extreme composition to the other, e.g. VSe-VSe_2.

ELECTRONIC CONFIGURATION

The order of orbital energies does not follow the same pattern in the filled and empty orbitals as is shown in Fig. 1.1. Since several sub-shells of the same quantum level are shielded to different degrees by the core of electrons beneath, their energies do not drop in a parallel manner with increase in atomic number.

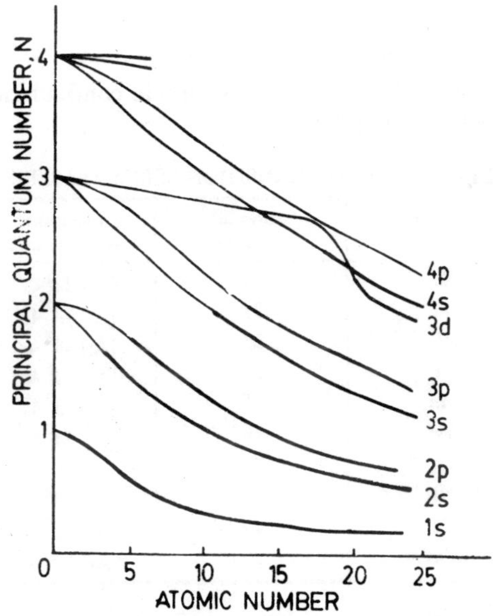

Fig. 1.1. Energy drop of orbitals as the orbitals are filled up.
Orbitals containing electrons ______
Empty orbital ______

From Fig. 1.1, it can be seen that 1*s*, 2*s*, 2*p* and levels occur in that sequence in all known atoms. Thus, from H to Ar electrons are filled up in that order. During the filling up of these orbitals the energies of higher (and as yet unfilled) orbitals are being altered by the shielding effect of these eighteen electrons. The 3*s* and 4*p* levels dropped steeply. Therefore, the two more electrons, added to argon to give the potassium and calcium atoms, enter the 4*s* orbitals which has fallen below the 3*d* orbitals. The nuclear charge has also been increased by two units. Due to increased penetration of 3*d* orbitals in the 4*s* orbitals, the effective nuclear for the 3*d* orbitals increases abrupty. As a result they drop well below the 4*p* orbitals and have almost the same energy as that of 4*s*-orbitals. Consequently, the next electron enters the 3*d* orbitals, and scandium has the configuration [Ar]$4s^2$ $3d^1$. This 3*d* electron screens the 4*p* orbitals more effectively rather than the remaining 3*d* orbitals. Thus, the additional electrons enter the 3*d* orbitals until they are filled up, in zinc, [Ar]$4s^2$ $3d^{10}$.

The extra stability of d^5 and d^{10} electronic configurations (due to higher exchange energies) forces the two elements, chromium and copper, to have configurations of $3d^5$ $4s^1$ and $3d^{10}$ $4s^1$ respectively instead of $3d^4$ $4s^2$ and $3d^9$ $4s^2$. The electronic configuration of the 3*d* series are listed in Table 1.1.

TABLE 1.1 : FIRST TRANSITION SERIES (3*d*-SERIES).

Element	Atomic number	Electronic configuration
Sc	21	[Ar] $3d^1$ $4s^2$
Ti	22	[Ar] $3d^2$ $4s^2$
V	23	[Ar] $3d^3$ $4s^2$
Cr	24	[Ar] $3d^5$ $4s^1$
Mn	25	[Ar] $3d^5$ $4s^2$
Fe	26	[Ar] $3d^6$ $4s^2$
Co	27	[Ar] $3d^7$ $4s^2$
Ni	28	[Ar] $3d^8$ $4s^2$
Cu	29	[Ar] $3d^{10}$ $4s^1$
Zn	30	[Ar] $3d^{10}$ $4s^2$

The electronic configuration of 4*d* and 5*d* series also follow the same sequence of events. The configuration of the last series (5*d*), however, is complicated by the still empty 4*f* level. Even less penetrating than *d*-orbitals are *f*-orbitals and are thus more effectively shielded by the inner electrons. As a result, the energy of the 4*f* level is significantly reduced only after the 4*s* orbitals has been filled up. Even so, the next electron enters the 5*d* levels to give lanthanum, before the seven 4*f* orbitals are filled, given a series of fourteen elements called the lanthanides. The filling of the 5*d* level is then completed followed by the 6*p*.

The configuration are determined by a net-effect of the following forces:

(a) Nucleus-electron attractions.
(b) Electron-electron repulsions.
(c) Shielding of electrons by each other.
(d) Exchange energy.
(e) Difference of energy between *d* and *s* orbitals.

The configurations of 4*d* and 5*d* series are listed in Table 1.2. and 1.3 respectively.

TABLE 1.2 : SECOND TRANSITION SERIES (4*d* SERIES).

Element	Atomic number	Electronic configuration
Y	39	[Kr] $4d^1\ 5s^2$
Zr	40	[Kr] $4d^2\ 5s^2$
Nb	41	[Kr] $4d^4\ 5s^1$
Mo	42	[Kr] $4d^5\ 5s^1$
Tc	43	[Kr] $4d^5\ 5s^2$
Ru	44	[Kr] $4d^7\ 5s^1$
Rh	45	[Kr] $4d^8\ 5s^1$
Pb	46	[Kr] $4d^{10}\ 5s^0$
Ag	47	[Kr] $4d^{10}\ 5s^1$
Cd	48	[Kr] $4d^{10}\ 5s^2$

TABLE 1.3 : THIRD TRANSITION SERIES (5*d* SERIES).

Element	Atomic number	Electronic configuration
La	57	[Xe] $5s^1\ 6s^2$
Hf	72	[Xe] $5d^2\ 6s^2$
Ta	73	[Xe] $5d^3\ 6s^2$
W	74	[Xe] $5d^4\ 6s^2$
Re	75	[Xe] $5d^5\ 6s^2$
Os	76	[Xe] $5d^6\ 6s^2$
Ir	77	[Xe] $5d^7\ 6s^2$
Pt	78	[Xe] $5d^{10}\ 6s^0$
Au	79	[Xe] $5d^{10}\ 6s^1$
Hg	80	[Xe] $5d^{10}\ 6s^2$

PERIODIC TRENDS

Atomic/Ionic Sizes

In going from left to right across a period the outermost electrons of consecutive elements all have the same value of *n*, the principal quantum number. With the increase in nuclear charge, there is a general decrease in atomic sizes. A large decrease is observed when the added electron is placed in a penetrating orbital, as illustrated by the large decrease in size between sodium (186 pm) and magnesium (160 pm). This is due to the fact that the added 3*s* electron penetrates the inner electron shells and also does not screen the other 3*s* electron from the increased nuclear charge.

In case of *d*-block elements, the added electron occupies an underlying, (*n*-1)*d* shell which provides some screening for the outermost *ns* electrons. Therefore, a small decrease in size is expected. However, the *d*-electrons do not penetrate the inner shells, and when sufficient numbers are present, interelectronic repulsions lead to a gradual increase in size. Thus at the end of the series there is a slight increase in size (Table. 1.4).

In a group, there is increase in atomic radium with increase in atomic number. This increase, however, is not as significant as in *s*- and *p*-block elements.

TABLE 1.4 : ATOMIC AND IONIC RADII OF 3*d*-ELEMENTS.

Radius (pm)	Sc	Ti	V	Cr	Mn	Fe	Co	Ni	Cu	Zn
Atomic Ionic⁺	144	132	122	117	117	116	117	115	117	124
M^{2+}	–	100	94	94	97	92	89	83	88	93
M^{2+}	89	81	78	76	79	79	75	74	–	–

⁺Ionic radii values are of high-spin complexes.

The sizes of elements following the lanthanides (lanthanum atomic no. 57 to lutecium atomic no. 71) are smaller than expected. This is associated with the filling up of 4*f*-orbitals which must be filled before the 5*d*-orbitals. The electrons in *f*-orbitals are not effective in screening other electrons from the nuclear charge. Therefore, there is gradual decrease in size of the lanthanide elements, which is termed as *lanthanide contraction.* Due to this phenomenon, the elements of second and third transition series are very close in size. Consequently, the properties of these elements are similar.

Ionic radii vary in a similar manner as the atomic radii (Table 1.4). The variation depends upon the filling up of the *d*-orbitals and the crystal-field stabilizations in the complexes.

Atomic Volumes and Densities

The atomic volumes of the transition elements are low as compared with those of Group I and Group II. This is due to the increased attraction (due to increased nuclear charge) which pulls the electrons inward. As a result densities of the transition metals are very high. In a series the densities increase across the period and reach a maximum value in Group VIII. This is explained in terms of small radii and close-packed structure of the elements.

Ionization Energies

Ionization energy generally increases with the increase in size of the elements. In the transition series the ionization energies increase along the period, though not in a regular manner. The increase is not as significant as for the *s*- and *p*-block elements in the same period.

TABLE 1.5 : IONIZATION ENERGIES OF THE ELEMENTS OF 3*d* SERIES.

Elements (3d series)	Ionization energies (kJ mol^{-1})		
	First	Second	Third
SC	632	1245	2560
Ti	659	1320	2721
V	650	1376	2873
Cr	652	1635	2994
Mn	716	1513	3258
Fe	762	1564	2963
Co	758	1647	3237
Ni	736	1756	3400
Cu	744	1961	3560
Zn	906	1736	3838

Table 1.5 shows the ionization energies of the 3*d* series. No regular increase in ionization energy is observed. This may be explained to be due to the extra shielding effect of the inner *d*-electrons.

The ionization energies of the transition metals are intermediate between those of the *s*-and *p*-block elements. This suggests that the transition metals are less electropositive than the *s*-block elements.

Successive (2nd, 3rd, etc.) ionization energies follow different trends. In the 3*d* series, second ionization energy increases almost regularly from Sc to Zn. Values for Cr and Cu are, however, much higher. This is due to their stable electronic configurations –$Cr^+(3d^5)$ and $Cu^+(3d^{10})$.

Magnetic Properties

The presence of unpaid electrons makes an atom or ion behave as a small magnet, giving rise to a magnetic moment. A substance in which there are atoms or ions with unpaired electrons thus behaves as though it contained a large number of minute magnets, and the application of an external magnetic field results in the lines of force being drawn through the substance. This phenomenon is known as paramagnetism and the substance is said to be paramagnetic.

When a magnetic field is applied to a substance in which all the electrons are paired, the induced magnetic moment in the compound acts in the opposite directions from the applied field. This effect is known as diamagnetism (Fig. 1.2). These effects can be used to measure the magnetic moment of a compound.

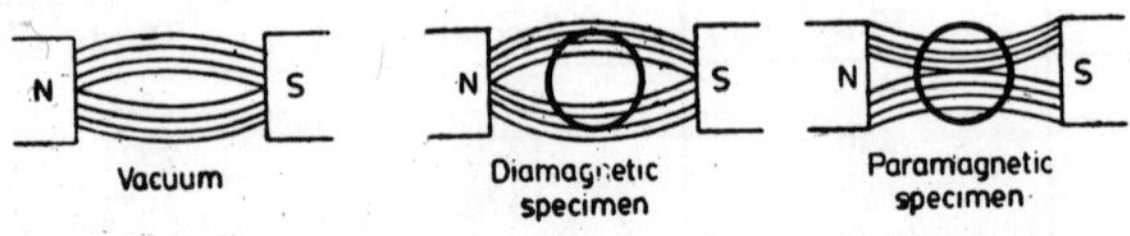

Fig. 1.2. Effect of diamagnetism and paramagnetism on the lines of force of a magnetic field.

The magnetic field produced by unpaired electrons is due to (a) their spin and (b) their orbital motion. The magnetic moment μ is given by

$$\mu_{S+L} = \sqrt{4S(S+1) + L(L+1)}$$

In situations where the orbital component can be neglected, the magnetic moment is given by the 'spin-only' formula

$$\mu_S = \sqrt{4S(S+1)}$$

or in terms of the number of unpaired electrons per atom, n:

$$\mu = \sqrt{n(n+2)}$$

The magnetic moment is measured in Bohr magnetons BM*. The spin-only values are 1.73 BM for the unpaired electron, 2.83 for two, 3.87 for the three, 4.90 for four, 5.92 for five, 6.93 for six and 7.94 for seven unpaired electrons.

TABLE 1.6 : MAGNETIC MOMENTS OF SOME IONS OF 3*d* SERIES.

Ion	Outer electronic configuration	No. of unpaired electrons *n*,	Magnetic moment (BM)	
			Calculated	Observed
Ti^{3+}	$3d^1$	1	1.73	1.75
Ti^{2+}	$3d^2$	2	2.84	2.76
V^{2+}	$3d^3$	3	3.87	3.86
Cr^{2+}	$3d^4$	4	4.90	4.80
Mn^{2+}	$3d^5$	5	5.92	5.96
Fe^{2+}	$3d^6$	4	4.90	5.0-5.5
Co^{2+}	$3d^7$	3	3.87	4.4-5.2
Ni^{2+}	$3d^8$	2	2.84	2.9-3.4
Cu^{2+}	$3d^9$	1	1.73	1.4-2.2

Another type of magnetic behaviour is *ferromagnetism,* which is very rare. It is a special case of paramagnetism and occurs in substances containing high proportion of atoms or ions with unpaired electrons. The individual ions align themselves parallel to the magnetic field and reinforce the magnetic moment due to individual ions. Ferromagnetic behaviour is found mainly among the metals alloys and oxides of transition elements.

Both paramagnetism and diamagnetism are temperature dependent. Ferromagnetic substances show normal paramagnetic behaviour above a certain temperature called the 'curie point'. In *antiferromagnetic* substances, the magnetic species align themselves in such a way that they cancel the effect of one another. This cancellation decreases with increase in temperature and the magnetic moment increases. Above a certain temperature, called the 'Neel

---

*(BM) = $\frac{eh}{4\pi mc}$

*where *e* is the electronic charge, *h* is the Planck's constant, *m* is the mass of the electron and c is the velocity of light.

point', the substance shows normal paramagnetism as there is no alignment.

Colour of the Transition Metal Ions

The colour of a substance is due to absorption of radiation is the visible region of the spectrum. The colour, of course, corresponds to the radiation which is not absorbed. When electron transitions occur from one energy level E_1 to a higher one E_2, there results absorption of radiation at frequency ν, corresponding to the Planck relationship $E_2 - E_1 = h$. The electrons can return to their original energy level by a series of smaller energy jumps, and hence there is a net absorption of radiation of frequency *v*. This would give rise to a single line in the absorption spectrum, but due to the various vibration in the ion or molecule an absorption band is produced.

The presence of electrons in the *d* level of a transition metal ion can also give rise to absorption. The different energy levels between which the electronic transitions are made are within the *d* level, and the spectra are referred to as *d-d* absorption spectra; they are responsible for the characteristic colours associated with transition metal ions, e.g. the blue of $[Cu(OH_2O)_6]^{2+}$ and the green of $[Ni(H_2O)_6]^{2+}$.

The spectra of transition-metal complexes are best explained on the basis of the crystal field theory. The splitting of the *d*-orbitals takes place under the influence of the surrounding ligands. The *d* orbitals are split by the surrounding ligands into the upper e_g and lower t_{2g} levels (Fig. 1.3).

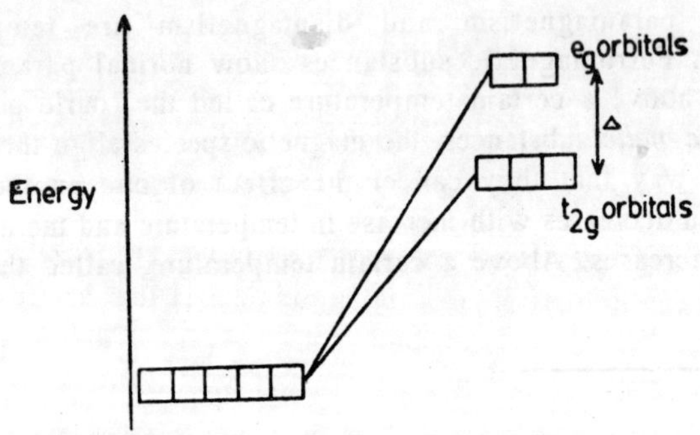

Fig. 1.3. Splitting of *d*-orbitals under the influence of a ligand.

For example, in $[Ti(H_2O)_6]^{3+}$ ion, the absorption spectrum can readily be accounted for in terms of the crystal field theory. In the ground state of the ion, the electron occupies the t_{2g} level. However, light brings about promotion of the electron to the e_g level, and the ion appears coloured. Maximum absorption occurs at the frequency corresponding to the mean value of Δ, the energy difference between the t_{2g} and e_g levels. This frequency is given the symbol ν_{max} and is usually measured in cm^{-1}. Often the absorption maximum is expressed as wavelength, in nanometres. The visible region of the spectrum extends from just below 14,000 cm^{-1} (714 nm)—red, to just above 25,000 cm^{-1} (400 nm)— violet end.

The spectrum of the $[Ti(H_2O)_6]^{3+}$ ion (Fig. 1.4) indicates that the light which is least absorbed is the red and violet. The hydrated

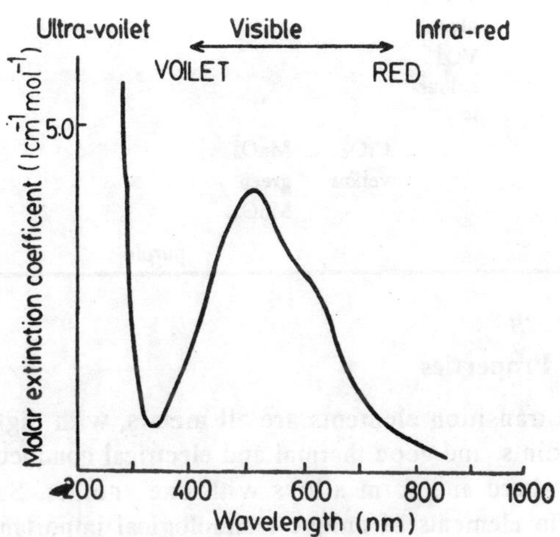

Fig. 1.4. Absorption spectrum of $[Ti(H_2O)_6]^{3+}$.

copper (II) ion has an absorption maxima in the far-red region. This is responsible for the characteristic blue colour of $CuSO_4$ solution.

Colours of some transition metal ions are listed in Table 1.7.

In molecules or complex ions it is often possible for electrons to make transitions from the orbital of one atom of the orbital of another, giving rise to electron transfer spectra or charge transfer specira. Usually, though not always, the electron transfer is from a ligand orbital to an orbitals of the central atom. It is responsible for the colour of oxyanions such as MnO^{4-}(purple), CrO_4^{2-} (yellow) and $Cr_2O_7^{2-}$ (orange).

TABLE 1.7 : COLOURS OF AQUO OR OXO IONS OF THE FIRST TRANSITION SERIES.

Sc	Ti	V	Cr	Mn	Fe	Co	Ni	Cu
Sc^{3+}	Ti^{2+}	V^{2+},	Cr^{2+}, Mn^{2+}, Fe^{2+},	Co2+, Ni^{2+}, Cu^{+},				
Color less	violet	violet	blue	pink	green	pink	green	colourl ess
	Ti^{3+},	V3+,	Cr^{3}, Cu^{2+}, Mn^{2+},	Fe3+,	Co3+,	Ni2+,		
	violet (green)	green	grey	red-brown	orange	blue	green	blue
		VO^{2+}, blue, VO_4^{3-}, colour-less						
			CrO_4^2 yellow	MnO_4^{2-}, green MnO4,				
					purple			

Metallic Properties

The transition elements are all metals, with high melting and boiling points, and good thermal and electrical conductives. They are generally hard and form alloys with one another. Such properties make them elements of unique technological importance.

Transition metals exhibit all the three types of structures–hexagonal close packed, face centred cubic and body central cubic. The coordination numbers are high and range from 8-12.

The valence electrons of the transition metal atoms are held rather loosely by the nuclii, and the atoms have no great affinity for more electrons. Bonding between metal atoms will thus be weak, and the pure elements may not exist as diatomic molecules. Instead stability is achieved by the sharing of valence electrons between many atoms. The metallic solid exists as a lattice arrangement of positive ions held together by delocalized electrons from the *ns* and (*n*–1) *d*-orbitals.

The lattice is so formed that each ion usually has eight either or twelve similar nearest-neighbour ions. The electrons are highly mobile. The presence of delocalized electrons confers a further property on the metallic lattice which distinguishes it from the lattice of an ionic-salt. This is the ability of the layers of ions to slide easily over one another on a 'cushion' of electrons. The metals are, therefore, malleable and ductile. Alloy formation is possible due to easy replacement of metal atoms by others of similar size without any appreciable distortion of the lattice.

Reactivity of Transition Elements

The transition metals exhibit high heats of sublimation, high ionization energies and low heats of hydration (solvation). Consequently, they show an increasing tendency to remain inactive or noble. The tendency to noble character is more pronounced in platinum and gold.

Oxidation States

One of the most striking features of the transition elements is that they exhibit a variety of oxidation states in their compounds. In contrast to this, the *s*-block elements show the group valency only and the p-black elements exhibit the group valency or a valency of two units less than the group valency. The following trends are generally seen in the oxidation states:

1. The maximum oxidation state in oxy, fluoro and chloro compounds corresponds to the total number of the *ns* and (*n*-1) *d* electrons.

   The oxidation state increases regularly (Sc^{3+} to Mn^{7+}). It reaches maximum value near the middle of each series (Mn^{7+}) and gradually decreases thereafter (Zn^{2+}). Stability of the maximum oxidation state decreases, but the colour increases.
2. In the oxidation states of II and III, the coordination number of ions is four or six.
3.

The stable oxyanions in the highest oxidation states are the tetrahedral MO_4^{n-} ions.

TABLE 1.8 : OXIDATION STATES OF TRANSITION ELEMENTS. FIRST TRANSITION SERIES.

Element	Sc	Ti	V	Cr	Mn	Fe	Co	Ni	Cu +1	Zn
Oxidation states	(+2)	+2	+2	+2	+2	+2	+2	+2	+2	+2
	+3	+3	+3	+3	(+3)	+3	+3	(+3)		
		+4	+4	(+4)	+4	(+4)	(+4)			
		+5								
				+6	(+6)	(+6)				
					+7					
Second transition series										
Element	Y	Zr	Nb	Mo	Tc	Ru	Rh	Pd	Ag	Cd
Oxidation states									1+	
						2+		2+	(2+)	2+
	3+		3+	3+		3+	3+	(3+)	(3+)	
		4+		4+	4+	4+	4+	4+		
			5+	5+		5+				
				6+	(6+)	6+	6+			
					7+	(7+)				
						(8+)				
Third transition series										
Element	La	Hf	Ta	W	Re	Os	Ir	pt	Au	Hg
Oxidation states									1+	1+
				(2+)		(2+)	(2+)	(2+)		2+
	3+			(3+)	(3+)	(3+)	3+	(3+)	3+	
		4+	(4+)	4+	4+	4+	4+	4+		
			5+	5+	5+	5+				
				6+	6+	6+	6+			
					7+					
					8+					

4. **Lower oxidation states are stabilized by π acid ligands like CO, NO, PR_3, etc.**

There is very small difference between the energies of electron in *ns* and (*n*-1) *d* orbitals. Thus both *ns* and (*n*-1) *d* orbitals participate in compound formation. Some of the oxidation states of the elements are listed in Table 1.8. The values reported in parentheses are either of uncommon or unstable oxidation states.

For a given transition element, the oxidation states differ by 1 rather than 2 as in the main-group (*p*-block) elements, where the variation state generally depends upon the inter-pair effect. The highest oxidation state is given by the total number (*N*) of *ns* (*n*-1) *d* electrons.

In the first transition series it is expected that the *ns* orbital would have two electrons and the ten transition elements one two ten

3*d* electrons. This is true except in case of Cu and Cr where one *s* electron moves into the *d* subshell, because of additional stability of exactly half-filled and completely filled *d*-orbitals.

Sc could, therefore, have an oxidation state of (II) if both *s* electrons were used. But Sc exhibits an oxidation state of (III) by utilizing two *s* and one *d* electrons. All other elements of the first transition series exhibit an oxidation state of (II) in addition to higher oxidation states.

Stability of Oxidation States

The relative stabilities of the different oxidation states are determined by several factors, e.g. the electronic configuration, the type of bounding, the stereochemistry, the lattice energy, the solvent and the solvation energy. The formation of cations is governed mainly by Fajan's rules. As the oxidation number increases, the covalent character increases, and the oxides become more acidic and halides are more easily hydrolyzed.

Lower Oxidation States

All the transition elements from compounds in lower oxidation states of 1, O, -1, etc. with π acid ligands. However, the compounds are not very stable except those of Cu(I) which are comparatively stable due to a d^{10} configuration.

Oxidation States II-IV

Well-defined, mainly ionic compounds are formed by all elements except Sc in the oxidation state II. The oxides are basic and the halides are ionic. With increase in the atomic number, the divalent state becomes more stable.

Oxidation state III is exhibited by all the elements except those of the zinc group. Sc(III) is very stable whereas Cu(III) is highly oxidising.

The IV Oxidation state is important for Ti and V only, while for the rest of the elements, it is found only in oxy and fluoro complexes.

Higher Oxidation States

The V, VI, and VII oxidation states are stable for vanadium, chromium and manganese respectively in the 3*d* series. Cr(VI) and

Mn(VII) are both highly oxidizing. Higher oxidation states become more stable for the *4d* and *5d* series, and are comparatively less oxidizing than the corresponding states of the *3d* series.

Complex Formation

A complex ion is one that contains more than one atom. Complex ions may contain either a metal ion or a nonmetal atom or ion as the central ion. The atom or molecule that attaches itself to a central atom or ion is known as a *ligand.*

The deep blue of a solution of copper sulphate containing ammonium hydroxide is caused by $\left[Cu(NH_3)_4\right]^{2+}$ ions. Hemoglobin, the red colouring matter in blood, is complex ion of iron; chlorophyll, the green colouring matter in plants, is a complex ion of magnesium; and vitamin B_{12}, a constituent of the vitamin-B group, is a complex ion of cobalt. Because water tends to hydrate ions readily, most ions in aqueous solution may be considered complex ions. Complex ions occur both in the solid state and in solution.

The transition elements are not the only metals whose ions form complex ions, but the tendency is very pronounced among this group of elements. Because of their small size, and comparatively high charge the transition metal ions exert strong force of attraction of the ligands. Vacant *d-orbitals* are used to form strong bonds with the ligands.

The stability of these complexes depends upon a number of factors such as the nature of the metal ion, its oxidation state, the nature of the donor atom, etc. metal ions with almost full *d*-orbitals, e.g. Cu^+, Ag^+, Ni^{2+}, etc. form stronger complexes with ligands of the type CO, NO, PR_3, etc. These ligands can accept the *d*-electrons of the metals in their vacant *d* or π orbitals through M–L π bonding. Transition metals with vacant d-orbitals form stronger complexes with N or O donor ligands. Such complexes are stabilized by the formation of strong σ bonds by electron donation from ligand donor atoms. For example, Co^{3+} forms a stable complex with ammonia — $\left[Co(NH_3)_6\right]^{3+}$.

Difference between *3d* and the Other Two Transition Series

Compared to the *3d* series, the elements of the other two (*4d* and *5d*) series differ as follows :

1. They are less abundant.
2. Due to lanthanide contraction, the atomic as well ionic radii of the *4d* and *5d* transition series are almost identical.
3. They form more stables II, III, etc. are less stable. For instance zirconium (III) is more strongly reducing than titanium (III). Simple aquo ions like M^{2+}(aq) and M^{3+} (aq), which are more common for *3d* series, do not exist for the *4d* and *5d* series.
4. Due to bigger size, they form compounds with higher coordination numbers, e.g. $[ZrF_7]^{3-}$.
5. *4d* and *5d* elements form complexes with large crystal field splitting energies. This results in low-spin complexes.
6. The elements of the other two series are much less reactive. This may be due to their higher ionization energies, lower hydration energies, higher sublimation energies, lower bond energies and lower lattice energies.

CHAPTER 2

Titanium, Zirconium and Hafnium–Group IV A

INTRODUCTION

The elements Ti, Zr and Hf differ from the other transition metals which follow them in their strong tendency to exhibit the group's maximum charge number, +4, almost to the exclusion of lower oxidation states; especially is this true of Zr and Hf. As in the latter transition subgroups, there is a distinct break in properties between the first member and the two heavier congeners. In fact Zr and Hf are so strikingly similar in their chemistry that the elements are extremely difficult to separate. Because of the effect of the lanthanide contraction which accompanies the filling of the 4f quantum level, the metallic radii and the estimated radii of the quadripositive ions are very close despite the difference of 2 in atomic number (Table 2.1).

Table 2.1 : ATOMIC PROPERTIES OF Ti, Zr and Hf

	Ti	Zr	Hf
Z	22	40	72
Electron configuration	[Ar] $3d^3 4s^2$	[Kr] $4d^2 5s^2$	[Xe] $4f^{14} 5d^2 6s^2$
$I(1)$/kJ mol^{-1}	658	670	530
$I(2)$/kJ mol^{-1}	1315	1345	1425
Metallic radius/pm	147	160	158
$^{r}M^{4+}$/pm	68	80	81

THE ELEMENTS

The elements are lustrous, silvery metals of high m.p. (Table 2.2). Titanium has a particularly low density for a transition metal and this, coupled with its strength and its resistance to corrosion, makes it of interest for the construction of supersonic aircraft. Ti and Zr have the h.c.p. structure at ordinary temperature but both have a high-temperature b.c.c. form; for Ti the transition temperature of the change is 1150 K. Hf has a structure very close to h.c.p. but six of the Hf-Hf distances are said to be about 2% longer than the other six. This metal also exists in a high-temperature b.c.c. modification.

TABLE 2.2 : PHYSICAL PROPERTIES OF Ti, Zr AND Hf

	Ti	Zr	Hf
ρ/g cm^{-3}	4.54	6.53	13.3
M.p./K	1950	2130	2470

The metals are extremely resistant to corrosion at ordinary temperatures, almost certainly because they are so easily rendered passive by the formation of an oxide layer. However, the metals react with most non-metals on heating; they combine with O_2, N_2, H_2, C, S and the halogens. Ti reacts particularly vigorously with nitrogen.

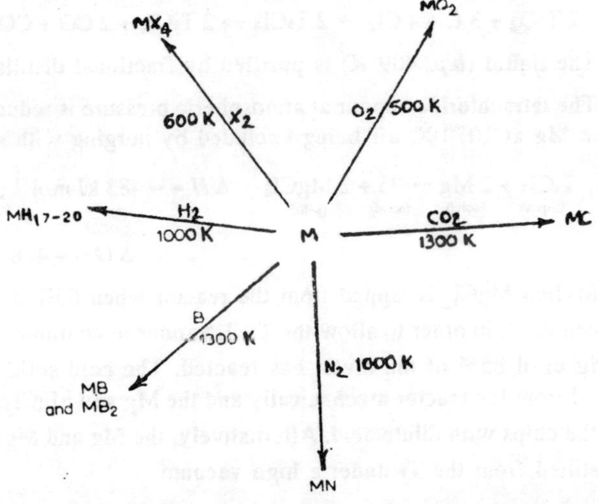

Fig. 2.1. Reactions of Ti, Zr and Hf (X = halogen).

The metals decompose steam on strong heating. Titanium dissolves in hot HCl to give $TiCl_3$, but resists attack by cold acids. Zirconium and hafnium dissolve even less readily in mineral acids, but all three metals dissolve in the presence of F^- ions, titanium giving salts of the Ti^{3+} ion, Zr and Hf giving zirconyl and hafnyl salts. The metals are remarkably resistant to corrosion by weakly acid solutions such as those of H_2S, SO_2, $FeCl_3$ (hot and cold) and even of H_2CrO_4. Hot and cold alkalis do not react with them.

EXTRACTION OF THE METALS

Titanium

The element (0.6% of the earth's crust) is abundant but difficult to extract. It is commonly associated with siliceous rocks, but the principal workable ores are ilmenite, $FeTiO_3$, and rutile, a tetragonal form of TiO_2. Reduction of TiO_2 with carbon is unsatisfactory because the very stable carbide is produced. The ease with which the metals combines with both oxygen and nitrogen at high temperatures makes other reduction methods difficult. The commercial production of titanium is by the metallothermic reduction of titanium tetrachloride.

In this process chlorine is passed over ilmenite or rutile heated to 1200 K with carbon, and the $TiCl_4$ vapour formed is condensed.

$$2\,TiO_2 + 3\,C + 4\,Cl_2 \rightarrow 2\,TiCl_4 \rightarrow 2\,TiCl_4 + 2\,CO + CO_2$$

The liquid (*b.p.* 409 K) is purified by fractional distillation.

The tetrachloride vapour at atmospheric pressure is reduced with molten Mg at 1070 K, air being excluded by purging with argon.

$$\underset{\text{(liquid)}}{TiCl_4} + \underset{\text{(solid)}}{2\,Mg} \rightarrow \underset{\text{(solid)}}{Ti} + \underset{\text{(solid)}}{2\,MgCl_2} \quad \Delta H = -483 \text{ kJ mol}^{-1};$$

$$\Delta G = -448 \text{ kJ mol}^{-1}$$

Molten $MgCl_2$ is tapped from the reactor when 60% of the Mg has been used, in order to allow the $TiCl_4$ vapour to continue to attack the Mg until 85% of the metal has reacted. The cold solid may be chipped from the reactor mechanically and the Mg and $MgCl_2$ leached from the chips with dilute acid. Alternatively, the Mg and $MgCl_2$ may be distilled from the Ti under a high vacuum.

The fragments of titanium are melted in a water-cooled copper crucible, by striking an arc between them and a compressed titanium-sponge cathode, in an atmosphere of argon.

Very pure titanium can be made by the iodine method in which TiI_4 vapour is decomposed on a hot wire.

The metal is unusual in igniting spontaneously in oxygen, at 2.4 MPa and room temperature, when a fresh surface is exposed by fracture. The reaction is self-propagating and leads to the complete combustion of massive pieces, through a superficial melting and rapid diffusion of the oxide into the metal, leaving a new surface of metal for oxidation. Similarly ZrO_2 dissolves in molten Zr, and this metal also behaves like Ti, but Mg, Al, Nb and Ta, whose oxides are not soluble in the metal, do not show the phenomenon.

The mechanical properties of titanium are comparable with those of steel; but it is more difficult to fabricate owing to the readiness with which it takes up, and is hardened and embrittled by, oxygen and nitrogen.

Zirconium

The principal ores of zirconium (0.025% of the lithosphere) are the silicate zircon, $ZrSiO_4$, and the oxide baddeleyite, ZrO_2. Treatment of these with carbon and chlorine at red heat gives crude $ZrCl_4$ which, after purification, can be reduced with Mg in a modification of the titanium process. For metal of higher purity the decomposition of ZrI_4 is used.

The metal is much softer than titanium. Its principal uses at present are in bullet-proof alloy steels and, because of its low cross-section for neutron capture, in alloys for cladding the metallic fuel elements used in some atomic reactors. Hafnium must be separated as completely as possible from zirconium which is to be used in reactors, because of the high cross-section for neutron capture of hafnium.

Hafnium

This element, predicted from atomic number sequence, was the first element to be discovered by X-ray methods. It was found in zirconium minerals which usually contain about 0.1% but occasionally up to 7% of hafnium. Because of its very close resemblance to zirconium, separation is difficult.

On a commercial scale a mixture of $ZrCl_4$ and $HfCl_4$ is dissolved in aqueous NH_4CNS, which causes some hydrolysis and some complex formation, and then agitated with hexone, which extracts the hafnium preferentially as $HfOCl_2$. The small amount of zirconium which enters the organic phase can be stripped from it with HCl. Finally the hafnium is recovered from the hexone with H_2SO_4, and the solvent is recycled.

OXIDATION STATES

The dipyridyl complexes Li [Ti (dipy)$_3$] 3.5 C_4H_8O and Ti (dipy)$_3$, which are both obtained by reducing $TiCl_4$ with lithium in solvent tetrahydrofuran in the presence of 2.2′-dipyridyl, are examples of titanium in the formal oxidation states -1 and 0 respectively. A compound of Zr°, Zr (dipy)$_3$, has been made similarly.

Compounds of Ti^{II} and Zr^{II} are known. The dihalides $TiCl_2$, $TiBr_2$, , TiI_2, $ZrCl_2$ and ZrI_2 have been made by reducing the tetrahalide vapours over the heated metals. These compounds are all oxidised by water, thus the + 2 states have no aqueous chemistry. An oxide of approximate composition TiO can be made by heating TiO_2 with Ti; it has a defect NaCl lattice. Titanium forms a cyclopentadienyl, $(\pi - C_5H_5)_2Ti$, and a cyclopentadienyl carbonyl, $(\pi - C_5H_5)_2Ti(CO)_2$, both Ti^{II} compounds.

The +3 state is represented in titanium by all four halides, the oxide Ti_2O_3, alums like $RbTi(SO_4)_2 \cdot 12H_2O$, halogen complexes like the hexafluorotitanate (III) ion, and complexes of $TiCl_3$ with oxygen and nitrogen donors. Aqueous solutions containing the violet $Ti(H_2O)_6^{3+}$ ion are useful reducing agents in volumetric analysis. They reduce Fe^{3+} to Fe^{2+}, chlorates and perchlorates to chlorides and aromatic nitro-compounds to amines; the redox potential for the Ti^{IV} aq/Ti^{III} aq couple is about +0.05 V. However, the solutions are rapidly oxidised in air and must be used in an atmosphere of N_2 or CO_2. The +3 state also occurs in trihalides of Zr and Hf, several of which have been prepared, but aquated Zr^{3+} and Hf^{3+} are not known.

The oxidation state +4 is dominate in the group; all compounds of the metals in lower oxidation states are easily oxidised to it.

HALIDES

Tetrafluorides result for the action of anhydrous HF on the chlorides :

$$TiCl_4 + 4\,HF \rightarrow TiF_4 + 4\,HCl$$

The white solids form stable complexes :

$$ZrF_4 + 2\,KF \rightarrow K_2\,Zr\,F_6$$

The TiF_6^{2-} ion is unstable to hydrolysis. It is converted rapidly to $TiOF_4^{2-}$. Surprisingly, some hexafluorozirconates, such as $K_2\,ZrF_6$, do not contain ZrF_6^{2-} ions, but instead ZrF_8 units formed by the sharing of F^- ions.

The tetrachlorides are made by passing chlorine over the dioxides heated with carbon. $TiCl_4$ is a colourless, strongly fuming liquid but $ZrCl_4$ and $HfCl_4$ are solids. The vapours are monomeric. Water hydrolyses them :

$$TiCl_4 + 2\,H_2O \rightarrow TiO_2 + 4\,HCl$$

$$ZrCl_4 + H_2O \rightarrow ZrOCl_2 + 2\,HCl$$

This oxide chloride crystallises from the solution as the octahydrate, $ZrOCl_2 \cdot 8\,H_2O$, which has a tetragonal structure containing $[Zr_4\,(OH)_8]^{8+}$ ions. Phase studies of zirconium tetrachloride-alkali-metal chloride systems indicate the formation of chlorozirconates $M_2^I\,ZrCl_6$. Ammonium chlorotitanate, $(NH_4)_2\,TiCl_6$, is precipitated when $NH_4\,Cl$ is added to a solution of $TiCl_4$ in concen- trated HCl.

The tetrabromides and tetraiodides are made by direct combination of the elements. Some are coloured; $TiBr_4$ is yellow and TiI_4 red-brown, in accordance with the position of the ligands in the spectrochemical series. They are solids of low *m.p.*; the crystals have a cubic lattice and contain tetrahedral molecules. Some bromo-complexes have been made, for instance $(NH_4)_2\,TiBr_6 \cdot 2\,H_2O$, but they are much less stable than the fluoro- compounds, Iodo-complexes are unknown; the stability falls rapidly with the more easily polarisable halogens.

Titanium trichloride, $TiCl_3$, is the most important trihalide. Its hexahy drate, like $CrCl_3 \cdot 6\,H_2O$ exhibits hydration isomerism. Rubidium and caesium chlorides give complexes containing $TiCl_5^{2-}$ ions.

The crystals, $M_2^I TiCl_5 \cdot H_2O$, are green but the solutions are violet. The colour change is associated with a change in the number of co-ordinated water molecules in the complex (cf. 34.4), TiF_3 is best made from titanium hydride and HF at 1000 K. It is a blue solid, stable to air, water and even concentrated H_2SO_4. Its magnetic susceptibility (1.75 B.M.) is appropriate to the Ti^{3+} ion with one d electron.

The trihalides ZrX_3 (X = Cl, Br or I), $HfBr_3$ and HfI_3 are known. They are oxidised by water to the +4 states, and therefore are without an aqueous solution chemistry.

Pure titanium dichloride is a dark-brown powder spontaneously inflammable in air. It is made by passing an electrode-less discharge through $TiCl_4$ mixed with hydrogen at low pressure. It sets free hydrogen from water. Chlorocomplexes $MTiCl_3$ and M_2TiCl_4 can be made by melting $TiCl_2$ with alkali-metal halides.

Titanium dibromide, which is made by reducing $TiBr_4$ with Ti, has been shown to have a CdI_2 structure. $ZrBr_2$ and ZrI_2 are obtained, mixed with the corresponding trihalides, when $ZrBr_4$ and ZrI_4 are reduced with hydrogen or the metal.

OXIDES

Both TiO_2 and ZrO_2 are manufactured for use as white pigments, TiO_2 from ilmenite by conversion to the sulphate, followed by hydrolysis (Fig. 2.2). The usual method is, however, the vapour-phase oxidation of $TiCl_4$, attained by passing the vapour with air through a flame produced by burning a hydrocarbon in an excess of oxygen.

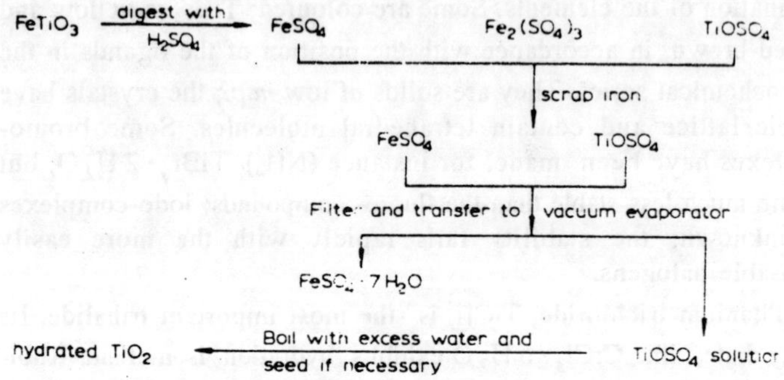

Fig. 2.2. Manufacture of TiO_2 from ilmenite.

The refractory oxide is white when cold but resembles SnO_2 in becoming yellow when hot. It has three crystalline forms, the tetragonal rutile, the slender tetragonal prisms of anatase and the flat plates of orthorhombic brookite. The pigment grades are either anatase or rutile. Rutile has 6 : 3 co-ordination and is isomorphous with cassiterite, SnO_2. In anatase, linear molecules of TiO_2 are present.

Zirconia (*m.p.* 2970 K) is used as a refractory and also as a pigment, mainly for white enamels. It is made from zircon (Fig. 2.3a) or baddeleyite (Fig. 2.3b).

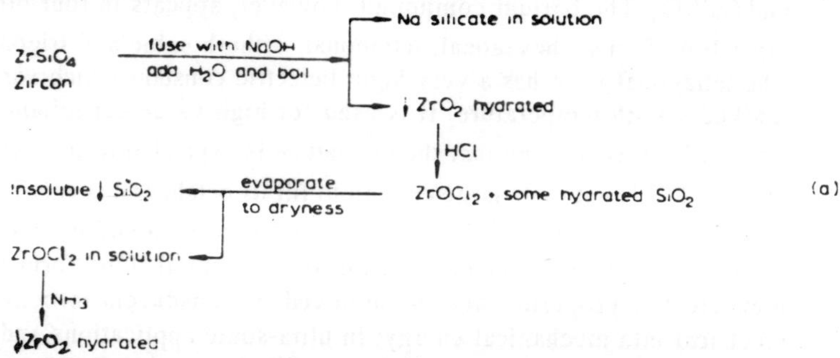

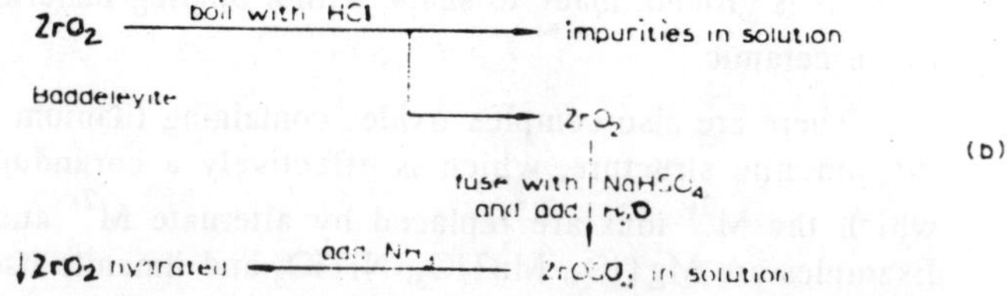

Fig. 2.3. Manufacture of zirconia, (a) from zircon, (b) from baddeleyite.

Hydrated oxides can be obtained by adding OH ions to solutions of the M^{IV} salts. They dissolve in acids. Hydrated TiO_2 also dissolves in concentrated aqueous alkalis to give hydrated compounds such as $M_2^1 TiO_3 \cdot n H_2O$, of unknown structure, but the hydrated ZrO_2 is almost insoluble in alkalis. When the hydrated oxides are strongly heated the MO_2 compounds which are produced are extremely resistant to attack by acids.

White TiO_2 can be converted to the dark-violet Ti_2O_3, which has the corundum structure, by strong heating in hydrogen at atmospheric pressure. It is not attacked by non-oxidising acids but it dissolves in some oxidising acids to give Ti^{IV} compounds. A non-stoi-chiometric TiO, with a defect-NaCl lattice, can be made by heating TiO_2 strongly with Ti.

Complex oxides

Most compounds formulated as titanates are complex oxides rather than salts. One of these, perovskite, $CaTiO_3$, gives its name to the perovskite structure of which other examples are $SrTiO_3$, $BaTiO_3$ and $CaZrO_3$. The barium compound, however, appears in four other crystalline forms : hexagonal, tetragonal, orthorhombic and trigonal. The tetragonal form has a very high dielectric constant which varies markedly with temperature. It is used for high value capacitances. The Ba^{2+} ion is so large that the O^{2-} lattice is expanded to the extent that the Ti^{4+} is not quite large enough to fill its octahedral space. Thus the Ti^{4+} ion is easily displaced within this space by a strong electric field, with a consequent polarisation of the crystal. One form has piezoelectric properties and is employed in transducers to convert electrical into mechanical energy, in ultra-sonic applications and in gramophone pick-ups. The material is made by heating $BaCO_3$ with TiO_2. It is ground, made to shape with a binding material, and fired like a ceramic.

There are also complex oxides containing titanium which have the ilmenite structure, which is effectively a corundum lattice in which the M^{3+} ions are replaced by alternate M^{2+} and M^{4+} ions. Examples are $MgTiO_3$, $MnTiO_3$, $NiTiO_3$ and ilmenite itself, $FeTiO_3$.

Other of the complex oxides of titanium have the spinel stricture examples are Mg_2TiO_4 and Zn_2TiO_4, made by heating the respective oxides with TiO_2, and $Mn^{II}Ti^{III}_2O_4$, produced when MnO_2 and TiO_2 are heated together at 1700 K. In Ba_2TiO_4, however, titanium is tetrahedrally coordinated to oxygen; the compound is a true titanate.

A number of complex oxides of zirconium are known; they are produced by strongly heating metal oxides, or nitrates, with ZrO_2. The calcium compound, $CaZrO_3$, is a perovskite ; but several of the $(M^{II})_2ZrO_4$compounds are known to be spinels.

BINARY COMPOUNDS WITH OTHER NON-METALS

Hard, refractory borides such as TiB and TiB_2 can be made by heating the metals strongly with boron. The carbides TiC and ZrC and be made similarly. These, and the mononitrides TiN and ZrN, also made by direct combination at high temperature, have the NaCl structure; the Ti atoms are in a cubic close-packed arrangement with the C or N atoms occupying octahedral spaces. Since the metals themselves have h.c.p. lattices these have not been simply expanded to accommodate the non-metal atoms.

Both Ti and Zr from nitrides M_3N_4. The titanium compound is made by the reaction between $TiBr_4$ and NH_2^- ions in liquid ammonia, and the zirconium compound by thermal decomposition of $Zr(NH_3)_4Cl_4$ obtained by the action of NH_3 on $ZrCl_4$. Both Ti_3N_4 and Zr_3N_4 lose nitrogen on heating and are converted to the mononitrides.

The three metals combine with sulphur to give disulphides, which are semi-conducting solids with metallic lustre. TiS_2 has been shown to have a structure similar to that of CdI_2.

AQUEOUS SOLUTION CHEMISTRY

Titanium (III) exists in aqueous solution as the blue-violet $Ti(H_2O)_6^{3+}$ ion. The hydrated ion is an acid of moderate strength ; for the equilibrium

$$Ti(H_2O)_6^{3+} + H_2O = Ti(H_2O)_5OH^{2+} + H_3O^+$$

the pK_a value is 3.9. The hexa-aquotitanium (III) ion occurs in some alums such as $RbTi(So_4)_2 \cdot 12H_2O$ and also in the violet form of $TiCl_3 \cdot 6H_2O$.

The Ti^{4+} ion is too strongly polarising to form simple aquo-complexes. Although salts of the Group IV A metals have formulae such as $TiOSO_4 \cdot H_2O$, $ZrO(NO_3)_2 \cdot 2H_2O$ and $Zr(C_2O_4)_2$ it is unlikely that MO^{2+} ions exist either in the aqueous solutions or in the crystal lattices. Solutions of titanium (IV) perchlorate contain $Ti(OH)_2(H_2O)_4^{2+}$ ions which are approximately octahedral, and solid $TiOSO_4 \cdot H_2O$ has been shown to contain zig-zag $(TiO)_n^{2n+}$ chains. The ions in zirconium (IV) crystals and their aqueous solutions are of even greater complexity. Solid $ZrOCl_2 \cdot 8H_2O$ contains

$[Zr_4(OH)_8(H_2O)_{16}]^{8+}$ ions in which the Zr atoms are linked in a distorted square by pairs of OH bridges. These ions are also thought to be the principal Zr^{IV} species present in molar aqueous $HClO_4$.

COMPLEXES

The halogen complexes are of interest. The compounds Na_3ZrF_7 and Na_3HfF_7 are reported to contain pentagonal bipyramidal MF_7^{3-} ions, but the ZrF_7^{3-} ion in $(NH_4)_3ZrF_7$ is a side-capped trigonal prism. Eight co-ordination of zirconium is also observed; the compound $Cu_2ZrF_8 \cdot 12H_2O$ contains ZrF_8^{4-} ions which are square antiprisms, and $Cu_3Zr_2F_{14} \cdot 16H_2O$ contains $Zr_2F_{14}^{6-}$ ions in which two antiprisms share an edge.

The tetrachlorides and tetrabromides of Ti and Zr act as Lewis acids and form adducts MX_4L and MX_4L_2 with oxygen donors such as carboxylates and also with various phosphorus and arsenic donors. The MX_4L adducts seem to be predominantly dimers, the metal atoms being connected by pairs of halogen bridges. As in the MX_4L_2 monomers there is approximately octahedral co-ordination of the metal atoms.

The tetrachloride and tetrabromides of all three metals react with β-diketones in inert solvents. The octahedral chloro-acetyl-acetonates $MCl_2(acac)_2$ have *cis*-chlorines. Zr and Hf also form compounds $MCl(acac)_3$ in which the metal atoms are 7-co-ordinate.

Complexes of bidentate ligands with donor atoms other than oxygen are also known. Orthophenylenebis (dimethylarsine) forms complexes MCl_4 (diars) and $MCl_4(diars)_2$, with all three tetra-chlorides

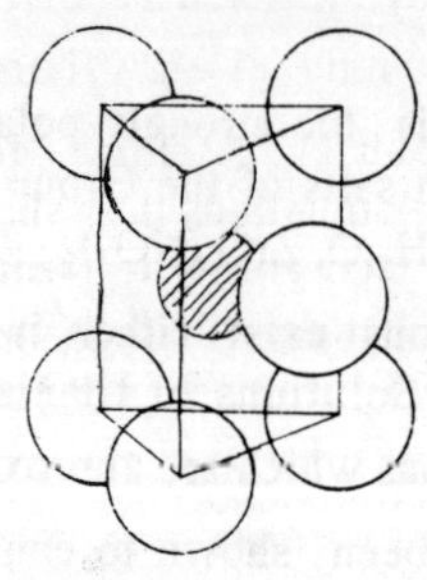

Fig. 2.4. ZrF_7^{3-} ion in $(NH_4)_3ZrF_7$.

The compound $TiCl_4$ (diars)$_2$ has a dodecahedrally co- ordinated Ti atom; it was an early example of the rare 8-co-ordination of a first-row d-block metal. The alkoxides of titanium have been extensively studied. They can be made by reactions of the type :

$$TiCl_4 + 4\,ROH + 4\,NH_3 \rightarrow Ti\,(OR)_4 + 4\,NH_4\,Cl$$

The compounds are used in heat-resisting paints in which they eventually hydrolyse to TiO_2. These alkoxides are liquids or solids which can be easily distilled or sublimed, and dissolve in non-polar organic solvents. Solid Ti $(OEt)_4$ has been shown to be a tetramer in which the four Ti atoms are coordinated approximately octahedrally to the oxygens of the OEt groups, some bridging and some terminal.

Titanium and zirconium both form volatile, anhydrous tetranitrates which are complex covalent compounds, not salts. Ti $(NO_3)_4$ is obtained when the hydrated nitrate is treated with $N_2\,O_5$, and Zr $(NO_3)_4$ by gently heating the adduct formed by the action of $N_2\,O_5$ on $ZrCl_4$. The metal atom in these molecules is co-ordinated to bidentate NO_3 groups through oxygen atoms arranged at the corners of a dodecahedron; the N atoms are thereby arranged in a slightly distorted tetrahedron. Zirconium also forms tetracraboxylates, a borohydride, Zr $(BH_4)_4$, and an anhydrous oxalate which are all covalent molecular complexes.

ORGANOMETALLIC COMPOUNDS

Unlike Ge, Sn and Pb in Group IVB, the metals of Group IV A do not form strong metal—carbon σ-bonds. The solid $C_6\,H_5$ Ti $(OPr^i)_3$ is obtained by treating Ti $(OPr^i)_4$ with phenyllithium. Although rapidly oxidised by air or water, it is thermally stable at room temperature in an inert atmosphere. $CH_3\,TiCl_3$ and Ti $(CH_2\,Ph)_4$ are other moderately stable compounds containing Ti—C σ-bonds.

Although the Group IV A metals do not form many sigma-bonded organometallic compounds they show general resemblance to the transition metals in their ability to form π-bonded organometallic compounds. Deep-red, diamagnetic bis (cyclopenta- dienyl) titanium dichloride, $(C_5\,H_5)_2\,TiCl_2$ is made by bromide and iodide are known; so are $(\pi\text{–}C_5\,H_5)_2\,ZrBr_2$ and $(\pi\text{–}C_5\,H_5)_2\,HfCl_2$; all are stable in air.

A tetracyclopentadienyl $(C_5H_5)_4$ Ti has been made which has the structure :

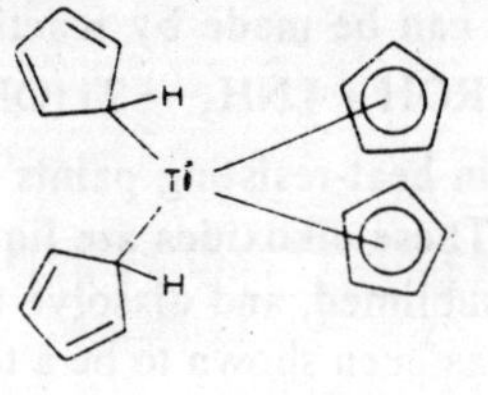

This molecule is fluxional in two senses. Firstly the σ-bonded C_5H_5 rings undergo 'ring-whizzing' and secondly the σ-bonded rings and the π-bonded rings rapidly change their roles. In consequence the ^1H n.m.r. spectrum at 298 K consists of a single broad absorption.

When $(\pi\text{-}C_5H_5)_2TiCl_2$ is reduced with sodium in tetrahydrofuran a compound of empirical formula $C_{10}H_{10}$ Ti is formed. This has two isomers, one of which has been shown to be a dimer with each Ti atom attached to a carbenelike C_5H_5 group, a $\pi\text{-}C_5H_5$ groups, and two bridging hydrogens. The compound is converted by carbon monoxide to $(\pi\text{-}C_5H_5)_2Ti(CO)_2$, the only well characterised carbonyl compound in this sub-group. Interestingly $C_{10}H_{10}$ Ti also absorbs molecular nitrogen; the product releases much of it as NH_3 on hydrolysis.

CHAPTER 3

Vanadium, Niobium and Tantalum – Group VA

INTRODUCTION

Vanadium, niobium and tantalum are typical transition metals. Their first ionisation energies are in the range 650 – 675 kj mol^{-1} (Table 3.1), they all crystallise with the b.c.c. lattice, they have high *m.p.* (Table 3.2) and they exhibit a wide variety of oxidation states in

TABLE 3.1 : ATOMIC PROPERTIES OF VANADIUM, NIOBIUM AND TANTALUM

	V	Nb	Ta
Z	23	41	73
Electron configuration	[Ar] $3d^3 4s^1$	[Kr] $4d^4 5s^1$	[Xe] $4f^{14} 5d^3 6s^2$
I (1)/kj mol^{-1}	651	654	675
Metallic radius/pm	134	146	146
$r_{M^{5+}}$/pm	56	70	73

TABLE 3.2 : PHYSICAL PROPERTIES AND ELECTRODE POTENTIALS OF VANADIUM, NIOBIUM AND TANTALUM

	V	Nb	Ta
p/g cm^{-3}	6.11	8.57	16.6
M.p./K	2190	2740	3270
E°, $M^{IV}{}_{aq}$/M/V	–1.5		
E°, $M^{V}{}_{aq}$/M/V		–0.6	–0.7

their compounds. The standard electrode potentials are extremely difficult to measure because the metals are so easily rendered passive that truly reversible electrodes cannot be prepared. Although the calculated values show the metals to be strong reducing agents, reaction does not occur with cold, dilute, non-oxidising acids. The metallic radius of Ta is very close to that of Nb, a consequence of the lanthanide contraction, and the two metals are therefore very similar in their chemistry, like the elements Zr and Hf which immediately precede them. Another consequence of the small metallic radius of Ta is that the metal has a high density, almost twice that of Nb.

Although unreactive at room temperate, the metals combine with most non-metals when heated. All three burn in oxygen to give the pentoxides. In the reactions of vanadium with halogens, the products depend upon the polarisability of the halide ion ; thus V reacts with F_2 at 570 K to give VF_5, with Cl_2 at 770 K to give VCl_4 and with Br_2 at 420 K to give VBr_3. However, most of the pentahalides of Nb and Ta can be made by direct combination ; even I_2 reacts with Nb at 570 K to give NbI_5. The metals also form nitrides, arsenides, carbides, silicides and borides by direct combination ; these compounds are mainly of interstitial type. The metals do not dissolve readily in cold mineral acids but they can be easily taken into solution as fluorocomplexes by a hot mixture of HF and HNO_3. Surprisingly, all three react with fused alkalis ; H_2 is liberated and oxoacid salts such as niobates are formed.

THE ELEMENTS

Vanadium

The element (0.02% of the lithosphere) is widely distributed — more than sixty vanadium minerals have been described — but there are few workable ores. Carnotite, $K(UO_2)VO_4 \cdot 1.5H_2O$, is a source of both uranium and vanadium ; vanadinite, $Pb_5(VO_4)_3Cl$, which is isomorphous with apatite, can also be worked up for the element, but the metal is being extracted to an increasing extent from residues obtained in iron and titanium production.

In the usual commercial process the vanadinite or vanadium residue is roasted with NaCl which converts the vanadium to sodium vanadate. This is leached out with water, and when the solution is acidified a red polyvanadate is precipitated. This can be fused at 950 K to give a commercial grade of V_2O_5, or dissolved in aqueous

Na_2CO_3 from which NH_4VO_3 is precipitated by NH_4^+ ions ; the ammonium vanadate is converted to a better grade of V_2O_5 by heating to 700 K.

Vanadium metal can be made by reducing V_2O_5 with calcium in a pressure vessel in the presence of a little iodine, some $CaCl_2$ being added to flux the lime which is formed :

$$V_2O_5 + 5\,Ca \rightarrow 5\,CaO + 2V$$

The metal is normally refined by electrolysis of a fused $NaCl/LiCl/VCl_2$ mixture, a process particularly effective for removing interstitial oxygen, nitrogen and hydrogen. Pure metal can also be made by the iodine process — the sublimation and thermal decomposition of VI_2.

The material ferrovanadium, which is made by aluminothermic reduction of V_2O_5 in the presence of iron, is manufactured for use in steel for highspeed tools. The alloy refines the grain and carbide structure of the steel, and improves its hardness at high temperature, by combining with the carbon which is present to form V_4C_3. Pure vanadium has little tensile strength and is a soft, ductile metal.

Niobium and tantalum

The elements are both rare, niobium being about 10^{-4} % and tantalum 10^{-5} % of the lithosphere. A mineral which is a mixed niobate and tantalate of iron and manganese, (Fe, Mn) (Nb, Ta, $O_3)_2$, is called columbite when it contains more Nb than Ta, otherwise tantalite. Another source of the metals is pyrochlore, which is a carbonatite — a mixed carbonate and silicate mineral — containing in this case some niobate and tantalate in addition. The columbite or pyrochlore is leached with aqueous HCl at 350 K for ten hours to take the Nb and Ta into solution. The metals are usually separated by a solvent-extraction procedure ; tantalum can be extracted first from weakly acidic solutions into methyl isobutyl ketone, the solution is then made strongly acidic and niobium is extracted by more of the same pure solvent. Niobium is recovered from the organic phase by agitation with de-ionised water, it is recovered by precipitation as $NbOF_3$, filtered, dried and calcined to Nb_2O_5. The metal can be obtained by reducing Nb_2O_5 with carbon or $NbCl_5$ with sodium. Tantalum is recovered from its organic phase similarly, and precipitated with KF as K_2TaF_7, which is dried and reduced to metai with sodium.

Niobium is used to inhibit intergranular corrosion in austenite steel, and as carbide in hard carbide tool compositions. An alloy with tin, of composition Nb_3Sn, is a particularly good superconducting material.

Tantalum is resistant to corrosion and for this reason is employed in both chemical research and plant construction. The element has minimal foreign body reactions in human tissue and finds a place in surgery. It is used in electrolytic rectifiers and in capacitors. These applications are possible because of the thin anodic film formed on the metal in oxoacid electrolytes.

Oxidation States

The lowest states are stabilised, as is usual, by π-acceptors such as CO. All three carbonylate anions $M(CO)_6^-$ are made by reduction of pentachlorides with sodium in diglyme under CO at 30 MPa. The vanadium compound $[Na(diglyme)_2]V(CO)_6$ can be oxidised to the dark-green, paramagnetic hexacarbonyl, which has an octahedral molecule and is an exception to the 18-electron rule.

The +1 state is also stabilised by ligands such as carbonyl and bipyridyr. The +2 state is unimportant, though there are well-established compounds such as double sulphates, containing the $V(H_2O)_6^2+$ ion. All the trihalides of vanadium are known, there are V^{III} alums and many anionic complexes such as $V(C_2O_4)_3^{3-}$ and $V(CN)_6^{3-}$. The +4 state is an important one in vanadium. It is

TABLE 3.3 : COMPOUNDS AND IONS REPRESENTATIVE OF THE OXIDATION STATES OF VANADIUM, NIOBIUM AND TANTALUM

	V	Nb	Ta
–1	$V(CO)_6^-$	$Nb(CO)_6^-$	$Ta(CO)_6^-$
0	$V(CO)_6$		
+1	$V(dipy)_3^+$	$(\pi\text{-}C_5H_5)Nb(CO)_4$	$(\pi\text{-}C_5H_5)Ta(CO)_4$
+2	$V(CN)_6^{4-}$		
+3	$V(NH_3)_6^{3+}$	$NbCl_3$	$TaBr_3$
+4	K_2VCl_6	$NbCl_4$	TaO_2
+5	$VOCl_3$	NbF_5	Na_3TaF_8

represented by halides such as VCl_4 and VBr_4, by vanadyl compounds such as $VOCL_2$ and $VOSO_4$ and by the dark-blue VO_2. Niobium and tantalum form some tetrahalides, of which the iodides are the best known, and also complexes of these with neutral ligands such as pyridine and diarsine.

As in the preceding group, however, the most common oxidation state is the maximum one, in this case +5. Niobium and tantalum form pentahalides with all four halogens and also have oxides M_2O_5. There is a great variety of halogen complexes as well as many types of vanadates, niobates and tantalates containing the metals in the +5 state. Vanadium has a fluoride VF_5, an oxide V_2O_5 and oxohalides such as $VOCl_3$ and $VOBr_3$. The relation between the four highest oxidation states of vanadium in aqueous solution is shown in Table 3.4.

TABLE 3.4 : VANADIUM COMPOUNDS SOLUTION

Vanadium charge number	+5	+4	+3	+2
Most common corresponding species and appropriate reducing agents	$VO_3^- \xrightarrow{(Fe^{2+})}$	$VO^{2+} \xrightarrow[\text{or } SO_2]{Sn^{2+},\ Ti^{3+}}$	$V^{3+} \xrightarrow[\text{or } Cr^{2+}]{Zn}$	V^{2+}
Colour in aqueous solution	Colourless	blue	green	violet
Redox potential		+1.0 V	+0.3 V	– 0.2 V
Typical compounds	NH_4VO_3	$VOCl_2$ $VOSO_4$	$V_2(SO_4)_3$	VSO_4
Typical complexes		$VO(SCN)_4^{2-}$	$V(NH_3)_6^{3+}$	$V(CN)_6^{4-}$

HALIDES

Vanadium forms the halides and oxohalides shown in Table 33.5. There is only one pentahalide, VF_5, made by direct combination at 600 K, but there are vanadyl trihalides in which the metal is formally in an oxidation state of +5 :

$$2VF_3 + O_2 \xrightarrow{\text{heat}} 2\,VOF_3 \text{ (white solid)}$$

$$V_2O_5 + \text{carbon} \xrightarrow[\text{inCl}_2]{\text{heat}} VOCl_3 \text{ (liquid, } b.p \text{ 400 K)}$$

$$V_2O_5 + \text{carbon} \xrightarrow[\text{in}Br_2]{\text{heat}} VOBr_3 \text{ (liquid, } b.p. \text{ 403 K)}$$

The brown liquid VCl_4 is made by direct combination at 800 K, and the limegreen solid VF_4 is obtained from it by the action of HF in a solvent freon.

The trihalides VBr_3 and VI_3 can be made from the elements but VCl_3 is obtained by treating the metal with HCl gas. The dark-coloured solids crystallise from water as green hexahydrates.

TABLE 3.5 : HALIDES AND OXOHALIDES OF VANADIUM

Charge number	Fluorides	Chalorides	Bromides	Iodides
+2	VF_2	VCl_2	VBr_2	VI_2
+3	VF_3	VCl_3	VBr_3	VI_3
	$VF_3 \cdot 6H_2O$	$VCl_3 \cdot 6H_2O$	$VBr_3 \cdot 6H_2O$	$VI_3 \cdot 6H_2O$
+4	VF_4	VCl_4		
	VOF_2	$VOCl_2$	$VOBr_2$	
+5	VF_5			
	VOF_3	$VOCl_3$	$VOBr_3$	
	VO_2F	VO_2Cl		

Yellow crystalline VF_3 is obtained by prolonged heating of VCl_3 with dry HF.

When VCl_3 is heated to 800 K in nitrogen, disproportionation occurs,

$$2\,VCl_3 \rightarrow VCl_2 + VCl_4$$

and the VCl_4 distils off, leaving VCl_2. VBr_3 and VI_3 both lose halogen on heating, giving brown VBr_2 and the violet VI_2 respectively. This last compound can be used in the preparation of pure vanadium by the iodine method. The blue solid VF_2 can be made by the action of HF gas on VCl_2 at 900 K.

The important halides of niobium and tantalum are the pentahalides, all made by direct combination, and the oxohalides such as $NbOCl_3$, $NbOBr_3$ and $TaOBr_3$. The pentachlorides and pentabromides are monomeric in the vapour phase, but X-ray diffraction studies on the crystals show them to be made up of M_2X_{10} units containing two distorted joined along one edge :

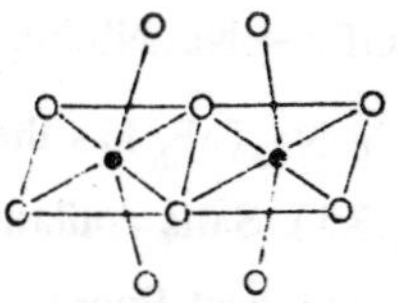

The oxohalides are usually made by heating the pentahalides in oxygen. They are less volatile than the halides themselves which can thus be separated from them by sublimation. The oxofluorides NbO_2F and TaO_2F are of structural interest, having the ReO_3 lattice with a random distribution of O and F in the anion positions.

With the exception of TaF_4, tetrahalides of Nb and Ta can be made by reduction of the pentahalides at high temperatures with Nb or Ta themselves, hydrogen or aluminium. They are diamagnetic, evidently because they contain metal — metal bonds.

Further reduction leads to a range of compounds containing Nb and Ta in formal oxidation states from +2 to +3. The majority of them are known to be cluster compounds, the most common cations in them being $M_6X_{12}^{n+}$ ions consisting of octahedra of metal atoms with halogens bridging all twelve edges. These cluster ions are joined in the crystals by doubly bridging halide ions. Thus the compound which is empirically Ta_2Cl_5 is formulated $[Ta_6Cl_{12}]Cl_{6/2}$, meaning that each cationic cluster is connected to three others by a total of six Cl^- ions. There is also an extensive solution chemistry of $M_6X_{12}^{n+}$ ions, where n = 2, 3 or 4. Salts such as $Nb_6Cl_{15} \cdot 7H_2O$ and $Ta_6Cl_{16} \cdot 7H_2O$ have been isolated from solution. The former contains the paramagnetic $Nb_6Cl_{12}^{3+}$ ion and the latter the diamagnetic $Ta_6Cl_{12}^{4+}$ ion. Anionic complexes containing metal clusters have also been obtained ; for example tetraethylammonium compounds of $Nb_6Cl_{18}^{n-}$, where n = 2, 3 or 4, can be made fróm aqueous solution containing $Nb_6Cl_{12}^{2+}$ ions and Cl^- ions.

The halides form numerous complexes. In particular the pentafluorides and the oxides fluorides combine with other metal fluorides :

$$VF_5 \xrightarrow{\text{KF in HF}} KVF_6$$

$$TaF_5 \rightarrow KTaF_6\text{ , }K_2TaF_7\text{ and }K_3TaF_8$$

$$NbOF_3 \rightarrow Na_3NbOF_6\text{ , }ZnNbOF_5 \cdot 6\,H_2O$$

The TaF_8^{3-} ion in Na_3TaF_8 has the form of a slightly distorted square antiprism (Fig 3.1). Salts containing NbF_7^{2-} have been made but not NbF_8^{3-}, and the highest fluoroanions which appear to exist in anhydrous HF are the MF_6^- species.

Many of the pentahalides form adducts with donor molecules. Examples are $NbF_5 \cdot OEt_2$, $TaF_5 \cdot SEt_2$, $NbCl_5 \cdot (C_5H_{11}N)_6$ and $TaCl_5 \cdot (C_5H_5N)_2$.

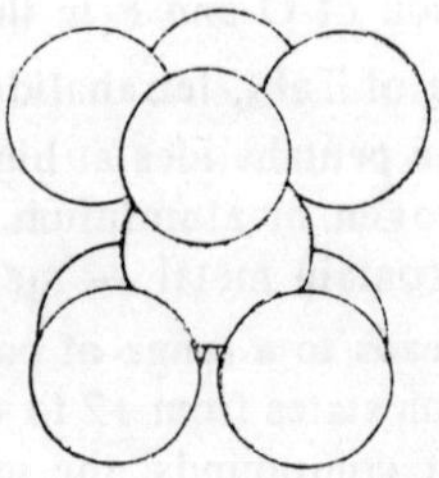

Fig. 3.1 Square antiprism. Structure of TaF_8^{3-} in Na_3TaF_8.

Orange-yellow V_2O_5 can be made by heating the metal in an excess of oxygen or by heating the "metavanadate" which has the empirical formula NH_4VO_3 :

$$2\,NH_4VO_3 \rightarrow V_2O_5 + 2\,NH_3 + H_2O$$

The oxide dissolves in strong alkalis to form orthovanadadates $M^3{}_1VO_4$ Some of these ($Na_3VO_4 \cdot 12\,H_2O$, $K_3VO_4 \cdot 6\,H_2O$) can be crystallised from solution at pH> 12. The addition of NH_4Cl to one of these solutions precipitates NH_4VO_3. When this is boiled for some time in 10% acetic acid it is converted to the golden $NH_4V_3O_8$, which is only one of a large number of polyvanadates. The decavanadates contain $V_{10}O_{28}^{6-}$ ions in which ten VO_6 octahedra share edges — these are isopolyanions similar to those found in Group VIA (34.6).

Various polyvanadate ions exist in alkaline aqueous solutions of V_2O_5. In solutions of concentration 10-100 mol m^{-3} of V_2O_5 it has been shown by spectrophotometric and potentiometric analysis that at pH < 2 the major species is VO_2^+, but between pH 2 and pH 6.5 the ions $H_2V_{10}O_{28}^{4-}$, $HV_{10}O_{28}^{5-}$ and $V_{10}O_{28}^{6-}$ are present in proportions depending on pH. At high pH the major species are $V_2O_7^{4-}$ and VO_3^{3-}. But at low concentration (< 0.1 mol m^{-3} of V_2O_5) the ions are mononuclear and the species which occur are VO_4^{3-}, HVO_4^{2-}, $H_2VO_4^-$, H_3VO_4 and VO_2^+, in proportions depending again on pH.

The pentoxide is converted by reduction into the other oxides :

$$\underset{\text{(blue-black)}}{VO_2} \xleftarrow[\text{(heat)}]{SO_2} V_2O_5 \xrightarrow[\text{(heat)}]{H} \underset{\text{(black)}}{V_2O_3} \xrightarrow[\text{at low pressure}]{\text{heat with V}} VO$$

Of these VO has the rock-salt structure, V_2O_3 the corundum structure and VO_2 the rutile structure, but the composition range associated with a particular lattice is always large, for example compositions as low as $VO_{1.35}$ are said to retain the corundum structure, and the rock-salt structure can persist through a range of composition from about $VO_{0.85}$ to $VO_{1.15}$.

It is probable that oxides with compositions between $VO_{1.66}$ and $VO_{1.86}$ contain phases which form the homologous series V_nO_{2n-1} with n = 3,4,5,6,7 or 8, based on the rutile structure. The reason V_2O_5 itself is such a good surface catalyst, as for example in the oxidation of SO_2 to SO_3, is almost certainly the ease with which vacant oxygen sites can be produced on heating. The structure of V_2O_5 is complicated ; double chains of distorted VO_5 bipyramids are linked into a layer structure by V—O—V bonds.

Hydrated oxides have been made. Rose-red $VO(OH)_2$ is made by concentrating, in an inert atmosphere, the acidic solution of a vanadate after reduction with SO_2. Black, crystalline $V_3O_5(OH)_4$ is obtained by reducing vanadic acid with zinc in the presence of concentrated NH_4Cl. A hydrated V^{III} oxide is precipitated by OH^- ions from vanadium (III) solutions but it is repidly oxidised in air.

The white oxides Nb_2O_5 and Ta_2O_5 are much more difficult to reduce than V_2O_5. Nb_2O_5 can be reduced with hydrogen at high temperature to NbO_2 which has a rutile-like structure, but similar treatment of Ta_2O_5 never reduces all the tantalum to the +4 state,

though an electrically conducting phase containing a few interstitial metal atoms can be produced.

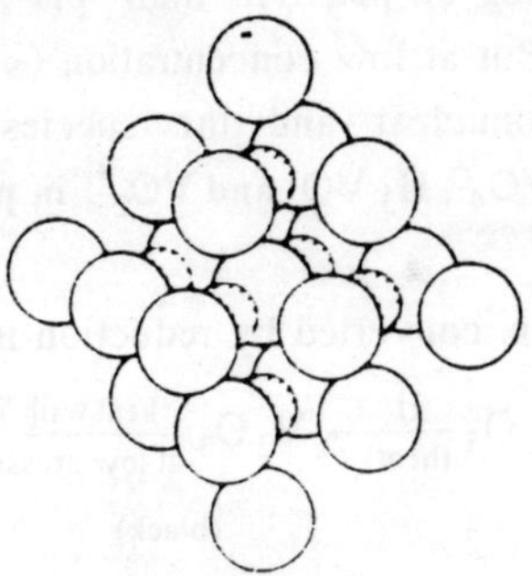

Fig 3.2 $Ta_6O_{19}^{8-}$ **ion.**

The pentoxides of niobium and tantalum are acidic. Fusion of Ta_2O_5 with caustic alkalis gives polytantalates of the form $M^I_8Ta_6O_{19}$. The anion $Ta_6O_{19}^{8-}$ can best be described as an octahedron of TaO_6 octahedra which share edges and have one oxygen atom common to all six (Fig. 3.2) Meta- and pyrotantalates, e.g., $Ca(TaO_3)_2$ and $Ca_2Ta_2O_7$, are also known. A compound $NaNbO_3$, with the perovskite structure, has been obtained by heating Nb_2O_5 with Na_2CO_3, and most of the compounds called niobates and tantalates are best considered as mixed metal oxides. However, the tetrahedral NbO_4^{3-} ion does exist in $ScNbO_4$.

BINARY COMPOUNDS WITH OTHER NON-METALS

Carbides

Very hard carbides result strongly heating the oxides with carbon. They are empirically MC and have the NaCl structure. Vanadium also forms V_4C_3 which has a defect structure, carbon atoms being missing from some of the lattice positions.

Nitrides

The elements combine with nitrogen at high temperatures to give the hard, very stable nitrides VN, NbN and TaN. They are not attacked by cold acids, but steam reacts with them at high temperatures to give the oxides and NH_3. The unstable higher nitrides Ta_3N_5 and VN_2 are made by heating the appropriate halides with ammonia.

Sulphides

The most stable sulphide of vanadium is V_2S_3 made by passing CS_2 over V_2O_5. It can be converted into V_2S_5 and VS.

$$V_2S_5 \xleftarrow[700\,K]{S} V_2S_3 \xrightarrow[1470\,K]{H_2} VS$$

Reduction of Ta_2O_5 with CS_2 at white heat gives TaS_2, the only sulphide of tantalum. NbS_2 is made by direct combination of the elements. The VS_4^{3-} ion is of interest because of its strong purple color which is like that of MnO_4^- ; presumably it also owes its color to charge-transfer absorption, as there are no partly filled d orbitals.

OXOACID SALTS OF VANADIUM AND NIOBIUM

The only oxoacid salts of importance are the sulphates of vanadium. A solid sulphate has not been obtained form a solution of V_2O_5, that is vanadium with charge number +5, in H_2SO_4, but reduction of the solution gives sulphates of vanadium with charge numbers +4, +3 and +2 respectively :

$$V^{V}{}_2O_5 \text{ in } H_2SO_4 \xrightarrow{SO_2} V^{IV}OSO_4 \xrightarrow[\text{sulphates}]{\text{alkali}} M^{I}{}_2SO_4 \cdot V^{IV}OSO_4 \cdot xH_2O$$

blue solution, dark blue double salts

$$\xrightarrow{Zn} V^{III}{}_2(SO_4)_3 \rightarrow \text{alums}$$

$$\xrightarrow[\text{reduction}]{\text{cathodic}} V^{II}SO_4 \cdot 7H_2O \text{ (isomorphous with } FeSO_4 \cdot 7H_2O)$$

A nitrate of niobium, $NbO(NO_3)_3$, has been made by treating $NbCl_5$ with N_2O_5 at 300 K.

$$NbCl_5 + 4N_2O_5 \rightarrow NbO(NO_3)_3 + 5NO_2Cl$$

ORGANOMETALLIC COMPOUNDS

As with the other transition metals, the π-bonded organometallic compounds are more numerous than the δ-bonded ones. The cyclopentadienyl compounds are of interest. Pale-green, paramagnetic $(\pi\text{-}C_5H_5)_2VCl_2$ is made by treating VCl_4 with sodium cyclopentadienide ; the dark-purple, paramagnetic $(\pi\text{-}C_5H_5)_2V$ is made by treating VCl_3 with the same reagent. A 'sandwich' structure has been proposed for the latter as a result of X-ray examination. The compound is soluble in most organic solvents and is very sensitive to air.

Reduction of $(\pi\text{-}C_5H_5)_2VCl_2$ with amalgamated zinc gives $(\pi\text{-}C_5H_5)_2VCl$; this reacts with phenyllithium to give $(\pi\text{-}C_5H_5)_2VC_6H_5$ in which the phenyl group is δ-bonded to the metal. From $NbBr_5$ the bis(cyclopentadienyl) niobium (V) tribromide $(\pi\text{-}C_5H_5)_2NbBr_3$ has been made by treatment with sodium cyclo- pentadienide.

The compound $(\pi\text{-}C_5H_5)_2TaH_3$ is of interest because its proton n.m.r. spectrum shows it to contain non-parallel C_5H_5 rings ; there is no evidence, however, that the bonding between rings and metal is fundamentally different from that in such compounds as ferrocene. The very reactive compound $(\pi\text{-}C_5H_5)_2NbH(C_2H_4)$, one of very few hydrido alkenes to have been made.

COMPLEXES

The metals of this sub-group form many complexes. In the oxidation state +5, high co-ordination numbers often occur, as in the TaF_8^{3-} ion already mentioned and in the NbF_7^{2-} ions which have the same shape as ZrF_7^{3-} (Fig.2.4). Many of the pentahalides react with oxygen and nitrogen donors :

$$NbCl_5 \xrightarrow{NHEt_2} NbCl_{3(NEt_2)_2} \cdot NHEt_2$$

$$TaBr_5 \xrightarrow[\text{inethanol}]{\text{acetylacetone}} TaBr_2(acac)(OEt)_2$$

$NbCl_5$ and $TaCl_5$ form 1 : 1 adducts with $POCl_3$ which are monomeric in benzene. In some organic solvents, $NbCl_5$ reacts with HCN to give the complex acid $HNbCl_5CN$; the triethylammonium salt of this acid, $Et_3NH\text{-}[NbCl_5CN]$, has been isolated.

Vanadium in the +4 state forms chelate complexes with salicylic acid, catechol and β-diketones. Examples are $M_2^1[VO(OC_6H_4CO_2)_2]$, $M_2^1[VO(C_6H_4O_2)_2]C_6H_6O_2 \cdot xH_2O$ and $VO(acac)_2$. In the last compound the vanadium atom is near the space centre of a square pyramid of oxygen atoms :

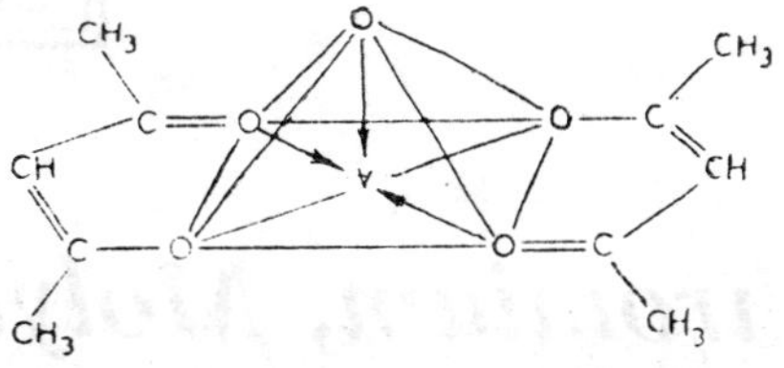

The 5-co-ordinate complexes of vanadium (IV) such as the acetylacetonate form adducts with nitrogen and phosphorus donors in which the metal becomes octahedrally co-ordinated. The introduction of a strong donor opposite to the V=O bond reduces the stretching frequency of that bond markedly, evidence that the metal is less able to accept electrons from the oxygen. Other vanadium (IV) complexes include adducts $VCl_4 L_2$ in which L is an alkylphosphine, as well as alkoxides $V_2 (OEt)_8$ and dialkuylamides $V(NR_2)_4$. The tetrahalides of Nb and Ta form adducts with nitrogen, oxygen and sulphur donors. Those with unidentate donors are usually of the $M X_4 L_2$ type and probably have *cis* octahedral configurations. Those with bidentate ligands such as $o\text{–}C_6 H_4 (AsMe_2)_2$ are believed to have a dodecahedral arrangement of atoms around the metal.

Vanadium (III) halides form 5-co-ordinate 1:2 adducts with some nitrogen and sulphur donors :

$$VCl_3 (SMe_2)_2, \quad VBr_3(NMe_3)_2$$

TLis oxidation state is also represented in chelate oxalate, β-diketone and catechol complexes :

$$V(C_2 O_4)_3^{3-}, \quad V(C_6 H_4 O_2)_3^{3-}, \quad V(acac)_3$$

The last compound is a dark-green solid, insoluble in water, but soluble in many organic solvents. Cyano and thiocyanato complexes are also known :

$$K_3 V(CN)_6 \quad \text{and} \quad K_3 V(CNS)_6$$

Vanadium(II) is represented by $K_4 V(CN)_6$. There double sulphates of the schonite type, such as $K_2 V(SO_4)_2 \cdot 6 H_2 O$, but these are lattice compounds rather than complexes.

CHAPTER 4

Chromium, Molybdenum and Tungsten—Group VIA

INTRODUCTION

Although both chromium and molybdenum have the ground state configuration d^5s^1 in contrast to d^4s^2 in tungsten, the main break in properties comes, as in the previous sub-groups of d-block elements, after Period 4. Chromium differs particularly from the other two in forming aquated bipositive and terpositive ions, which are similar to those formed by the other transition metals of Period 4.

As a consequence of the lanthanide contraction, the metallic radii of molybdenum and tungsten are the same, as are the radii of their 4+ cations (Table 4.1). The first ionisation energy of chromium is rather low for the metal's place in the Periodic Table but those of Mo and W are unremarkable.

TABLE 4.1 : ATOMIC PROPERTIES OF CHROMIUM, MOLYBDENUM AND TUNGSTEN

	Cr	Mo	W
Z	24	42	74
Electron configuration	[Ar] $3d^54s^1$	[Kr] $4d^55s^1$	[Xe] $4f^{14}5d^46s^2$
$I(1)$/kj mol^{-1}	653	692	770
Metallic radius/pm	127	139	139
Ionic radius M^{3+} /pm	64		
M^{4+} /pm	55	68	68

The three metals are silvery white ; they have b.c.c. structures at room temperature and are rather soft when carefully purified. The *m.p.* are high (Table 4.2), that of each metal being at the peak of its

TABLE 4.2 : PHYSICAL PROPERTIES OF CHROMIUM, MOLYBDENUM AND TUNGSTEN

	Cr	Mo	W
p/g cm^{-3}	7.14	10.8	19.3
M.p/K	2180	2890	3680

respective transition series ; thus tungsten has the highest *m.p.* of any metal and is consequently used for electric light filaments. Judged potentiometrically, chromium is a strong reducing agent (E°, $Cr^{3+}/Cr = -0.74$ V) but reactivity is inhibited by the formation of a touch oxide coating. The resistance of the metal to corrosion makes it a valuable protective plating on steel.

The metals are rather unreactive at laboratory temperature. Neither Mo nor W is easily dissolved by non-oxidising acids, though Mo is attacked readily by a mixture of concentrated HNO_3 and HF. Chromium dissolves in cold HCl but is rendered passive by HNO_3. The metals are much more reactive at high temperatures ; they combine with oxygen, sulphur, nitrogen, carbon, silicon and boron ; with the last four elements most of the products are interstitial compounds. With oxygen, Mo and W combine directly to form the trioxides MoO_3 and WO_3 at red heat, but Cr gives Cr_2O_3 in this way. Fluorine attacks the metals in the cold but the other halogens do so only on strong heating.

THE ELEMENTS

Chromium

The element (0.2% lithosphere) is produced from chromite, $FeCr_2O_4$, which is a spinel. Direct reduction of this compound with carbon in an electric furnace yields ferrochrome, an iron — chromium alloy used in making stainless steel (12 — 26% Cr) and tool steels (3 — 6% Cr). Chromium itself can be obtained 99.8% pure by the electrolysis of aqueous chrome alum prepared from a solution of ferrochrome in sulphuric acid. The metal can be electroplated on to other surfaces to provide decoration, but steel is best plated first with either nickel or copper on which the Cr is deposited, because the decorative plating is rather porous. The purest chromium is made by the decomposition of chromium iodide, produced from the elements at 1050 K, on a filament heated to 1500 K.

The metal dissolves slowly in cold HCl and H_2SO_4, rapidly in the same acids when they are hot, but not at all in HNO_3. It reacts with chlorine at 900 K and bromine at red heat, with oxygen at 2300 K and with steam at red heat :

$$2\,Cr + 3\,X_2 \rightarrow 2\,CrX_3 \text{ (X = Cl or Br)}$$

$$4\,Cr + 3\,O_2 \rightarrow 2\,Cr_2O_3$$

$$2\,Cr + 3\,H_2O \rightarrow Cr_2O_3 + 3H_2$$

Molybdenum

The element (1.5×10^{-4}% of the lithosphere) is extracted from molybdenite, MoS_2, a substance which, because of its laminar structure, is also used as a lubricant, particularly for conditions of high temperature and high pressure. The concentration of MoS_2 in the ore is low, usually about 0.3%, but it can be recovered as a 90% concentrate by a flotation process. Roasting converts the MoS_2 to MoO_3, which can be either purified by sublimation or converted to ammonium molybdate by ammonia treatment. Molybdenum powder is obtained by reducing either the MoO_3 or the ammonium molybdate with hydrogen at 1000 K. The metal in its massive form is produced by sintering the powder in hydrogen at about 2500 K and 2 MPa. It is used for toughening steel and for supporting tungsten filaments in electric lamps. Annual world production is about 60 thousand tons.

Molybdenum reacts with fluorine at room temperature, with Cl_2 and Br_2 above 600 K, but not with iodine except under pressure. Thus the iodine method cannot be used, as in the case of chromium, for the production of the purest molybdenum, which must be made by the hydrogen reduction of carefully prepared MoO_3. The metal is not attacked by HCl, HF or H_2SO_4 ; it reacts initially with HNO_3, but quickly becomes passive. It is, however, attacked rapidly by fused oxidising agents such as Na_2O_2, $KClO_3$ and KNO_3.

Tungsten

The element (1.6×10^{-4}% of the lithosphere) is made principally from scheelite, $CaWO_4$, or from wolframite, a mixed crystal of $FeWO_4$ and $MnWO_4$. About 90% of the world's production of the metal (15 thousand tons per annum) is used in the manufacture of ferrous alloys, mainly tool steels (18–20% W). For this purpose ferrotungsten is made by reducing high grade concentrates of $FeWO_4$, low in manganese, with coke in an electric furnace. Tungsten for electric lamps is made from $CaWO_4$ which can be converted by HCl to tungstic acid subsequently purified by dissolving in aqueous alkali and reprecipitating. The acid is calcined to WO_3 which is reduced with hydrogen to powdered metal that can be converted to massive metal in the same way as molybdenum. Industrially, the most important compound of tungsten is its carbide, produced by direct combination at high temperature. Sintered with cobalt powder, it produces a material which is almost as hard as diamond and is applied where great resistance to wear is important, as in rock-drilling tools.

The reactivity of tungsten is very similar to that of molybdenum, but whereas Mo is converted to $MoCl_5$ by chlorination and to $MoBr_3$ by bromination, the products in the case of tungsten are the hexahalides.

OXIDATION STATES

For Cr, as for Ti and V in the earlier groups, the highest oxidation state is that corresponding to the periodic group number, but Cr^{VI} compounds differ from those of Ti^{IV} and V^{V} in having strong oxidising power, for example $E°$, $Cr_2O_7^{2-}/Cr^{3+}$ is + 1.33 V. For the higher congeners Mo and W the +6 state is much more stable to reduction, however (Fig. 4.1). This relation between the heavier d-block elements and those in the first row is repeated in the later transition groups also.

Cr, Mo and W all form diamagnetic hexacarbonyls in which the formal oxidation state of the metal is zero, and some carbonylate anions are known which contain the metals in negative oxidation states. For chromium the Cr^{3+} ion has an extensive aqueous chemistry, but the Cr^{2+} ion is a strong reducing agent ($E°$, Cr^{3+}/Cr^{2+} = —0.4 V). Mo and W do not form simple hydrated cations in aqueous solution.

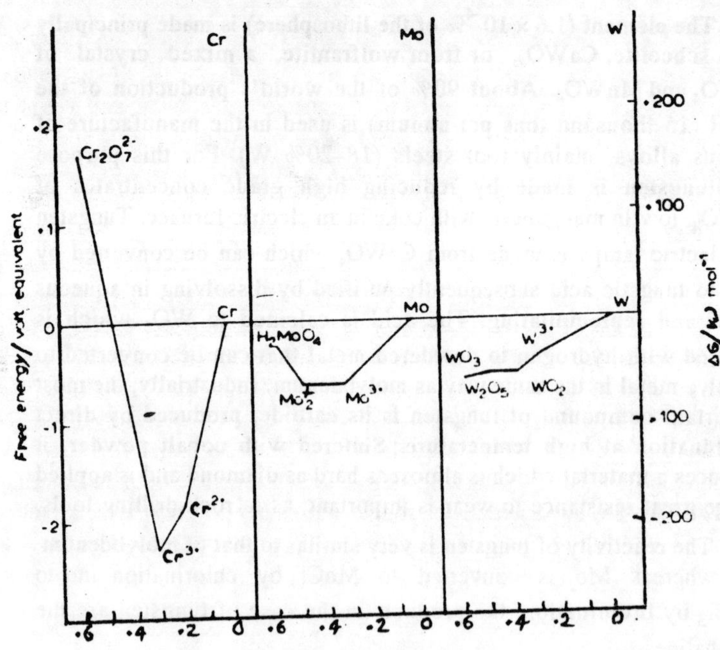

Fig. 4.1. Free energies of the oxidation state of Group VIA metals relative to the metals in aqueous solution at pH = 0.

Representative compounds of the various oxidation states are shown in Table 4.3. As is general in this part of the Periodic Table, the lowest oxidation states are stabilised by π -acceptors such as

Table 4.3 : COMPOUNDS AND IONS REPRESENTING THE OXIDATION STATES OF CHROMIUM, MOLYBDENUM AND TUNGSTEN

Oxidation state	Cr	Mo	W
—2	$Cr(CO)_5^{2-}$	$Mo(CO)_5^{2-}$	
—1	$Cr_2(CO)_{10}^{2-}$		
0	$Cr(C_6H_6)_2$	$Mo(CO)_3py_3$	$W(CO)_6$
+ 1	$Cr(dipy)_3^+$	$(C_6H_6)_2Mo^+$	
+ 2	$CrCl_2$	Mo_6Cl_{12}	$W_6Cl_8^{4+}$
+ 3	$Cr_2(SO_4)_3$	$Mo(CN)_6^{3-}$	$K_3W_2Cl_9$
+ 4	Ba_2CrO_4	MoS_2	WO_2
+ 5	CrF_5	$Mo(CN)_8^{3-}$	WF_6^-
+ 6	CrO_4^{2-}	MoO_2Cl_2	WF_8^{2-}

CO. the intermediate states by ligands such as CN^-, Cl^-, Br^- and H_2O, and the highest states by the hard bases F^- and O^{2-}, though in tungsten the +6 state is of particular importance and is exemplified by the hexahalides WCl_6 and WBr_6 as well as WF_6, the oxide WO_3 and its corresponding acid and salts. Although the +6 state is also important in molybdenum, the +5 state is stabilised by many common ligands ; the highest chloride of Mo, for example, is $MoCl_5$.

HALIDES

The halides of Cr, Mo and W are shown in Table 4.4, which illustrates the usual trends to be found in the halides of d-block congeners. The metals of periods 5 and 6 form more of the higher halides than does the Period 4 element. Thus Mo and W, like the Group IV and Group V metals which precede them, form several halides in which all the outer d and s electrons are used in bonding. The +6 state is particularly strongly represented in tungsten, a fact of some interest because this is an unusual oxidation state for metals to exhibit in their chlorides and bromides.

Table 4.4 : HALIDES OF CHROMIUM, MOLYBDENUM AND TUNGSTEN

Charge numbers	Cr		Mo		W	
+ 2	CrF_2,	$CrCl_2$	Mo_6Cl_{12} *			W_6Cl_{12}
	$CrBr_2$,	CrI_2	Mo_6Br_{12},	Mo_6I_{12}	W_6Br_{12},	W_6I_{12}
+ 3	CrF_3,	$CrCl_3$	MoF_3,	$MoCl_3$		WCl_3
	$CrBr_3$,	CrI_3	$MoBr_3$,	MoI_3	WBr_3,	WI_3
+ 4	CrF_4		MoF_4,	$MoCl_4$	WF_4,	WCl_4
				$MoBr_4$	WBr_4,	WI_4
+ 5	CrF_5		MoF_5,	Mo_2Cl_{10}		WCl_5
					WBr_5	
+ 6	CrF_6		MoF_6		WF_6,	WCl_6
					WBr_6	

* The brown dichloride, $MoCl_2$, which is distinct from yellow Mo_6Cl_{12}, is also known.

The chromium (II) halides are made by the action of hydrogen halides on the metal at 1000 K (CrF_2, $CrCl_2$ and $CrBr_2$) or by direct combination of the elements (CrI_2). The d^4 configuration gives rise to Jahan — Teller distortion ; in the CrF_2 crystal for example there is a tetragonal arrangement of F^- ions around the Cr^{2+} ion, with four of the fluoride ions at 200 pm and two at 243 pm. Although a true dihalide of molybdenum, $MoCl_2$, can be made by the action of chlorine on $Mo(CO)_6$, the more common halides containing Mo^{II} and W^{II} are cluster compounds of formula $(M_6X_8)^{4+}(X^-)_4$. The complex cation $Mo_6Cl_8{}^{4+}$ has been shown to be cubic, with Cl atoms at the eight corners and Mo atoms at the six face-centres (Fig. 4.2). The $M_6X_8{}^{4-}$ ion can co-ordinate six electron-pair donors, one to each metal atom along a four fold axis of the octahedral cluster.

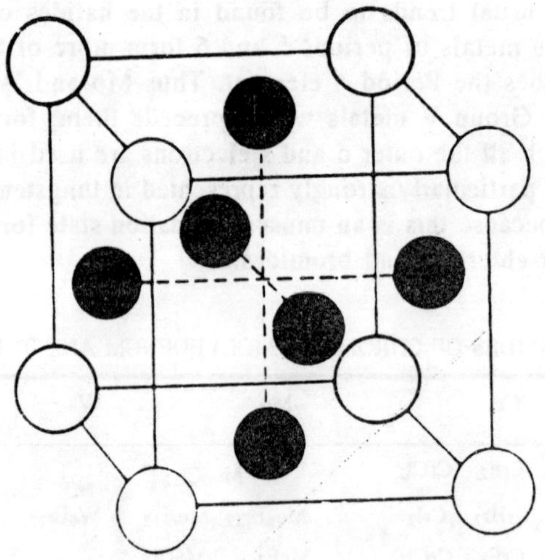

Fig. 4.2. Metal cluster ion $Mo_6Cl_8{}^{4+}$.

In Mo_6Cl_{12} each $[Mo_6Cl_8]^{4+}$ ion is co-ordinated thus to four bridging Cl^- ions and two non-bridging ones. This compound is prepared by the reduction of $MoCl_4$ with Mo at about 900 K, while W_6Cl_{12} and W_6Br_{12} can be made by heating the pentahalides with aluminium.

Red-violet $CrCl_3$ can be made by heating the metal with chlorine or the oxide Cr_2O_3 with CCl_4. The colour of an aqueous solution varies with temperature and chloride-ion concentration. In a cold, dilute solution the octahedral $Cr(H_2O)_6{}^{3+}$ ion is violet, but as the temperature is raised and the Cl^- concentration is increased, the colour changes to green because of the reaction :

$$Cr(H_2O)_6^{3+} + Cl^- \rightarrow CrCl(H_2O)_5^{2+} + H_2O$$

Three isomers of $Cr(H_2O)_6Cl_3$ can be crystallised from aqueous solutions depending on the conditions. The tribromide, however, has only two hydration isomers and the tri-iodide has only one form.

Of the trihalides of Mo and W, those of most interest are the brown solid MoF_3, made by the reduction of MoF_6 with Mo at about 700 K, and WCl_3 which is made by treating W_6Cl_{12} at 370 K with Cl_2. It is really a cluster compound $(W_6Cl_{12})Cl_6$ in which the $W_6Cl_{12}^{6+}$ ion consists of an octahedron of W atoms bridged along every edge by Cl atoms.

The higher fluroides of chromium are made by direct combination of the elements, CrF_4 at about 600 K, CrF_5 between 650 K and 750 K, and CrF_6 at 700 K in a bomb at 20 MPa.

The tetrahalides of Mo and W are not, in general, easy to obtain pure ; the fluorides can be made by reducing the hexafluorides with benzene at about 380 K, and WCl_4 results from the reduction of WCl_6 with aluminium in a thermal gradient. It disproportionates to W_6Cl_{12} and WCl_5 at about 750 K. The most stable pentahalides are Mo_2Cl_{10} and WBr_5, both made by direct combination of the elements at high pressure.

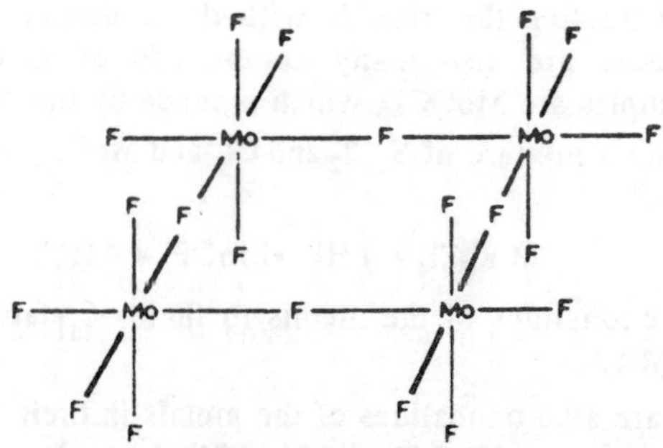

Fig. 4.3. Tetramer of molybdenum (V) fluoride.

The dark-green crystals of molybdenum(V) chloride contain Mo_2Cl_{10} molecules similar in structure to Nb_2Cl_{10} (3.4). The compound is paramagnetic, with $\mu_{eff} = 1.6\ \mu_B$ at 293 K, indicating a very slight coupling of the electron spins of the two metal atoms. In the vapour and in solvents such as benzene the compound exists as $MoCl_5$. The pentafluoride, MoF_5, a yellow solid, is made by reducing MoF_6 with $Mo(CO)_6$:

$$5\ MoF_6 + Mo(CO)_6 \rightarrow 6\ MoF_5 + 6\ CO$$

It has a tetrameric structural unit (Fig. 4.3) in which the arrangement of F atoms around the metal atoms is, however, only approximately octahedral.

The hexafluorides of Mo and W are colourless, volatile, diamagnetic compounds, both made by direct fluorination of the metal. The chloride WCl_6 and the bromide WBr_6 can also be made by direct halogenation of the metals ; they are dark-blue solids. These covalent hexahalides are all readily hydrolysed by water.

OXOHALIDES

The three metals form many oxohalides of the type MO_2X_2. In the case of Cr and Mo, X can be F, Cl or Br, in the case of W, X can be Cl, Br or I. Chromyl chloride, CrO_2Cl_2, familiar as a distinguishing test for a chloride, is the yellow distillate obtained when a chloride is heated with $K_2Cr_2O_7$ and concentrated H_2SO_4 :

$$Cr_2O_7^{2-} + 4\ Cl^- + 6\ H_2SO_4 \rightarrow 6\ HSO_4^- + 3\ H_2O + 2\ CrO_2Cl_2$$

The compounds are mostly yellow or orange liquids or low-melting solids ; those of molybdenum and tungsten can usually be made by treating the trioxide with the hydrogen halide or the halogen. There are also many compounds of general formula MOX_4. Examples are $MoOCl_4$, which is made by the chlorination of MoO_2Cl_2 with a mixture of S_2Cl_2 and Cl_2 and $MoOF_4$ which is made from it :

$$MoOCl_4 + 4\ HF \rightarrow MoOF_4 + 4\ HCl$$

These oxohalides of the metals in their +6 states form many complexes (4.10).

These are also oxohalides of the metals in their +6 states, and compounds such as $CrOCl_3$ and $MoOCl_3$ have been made. The chromium compound disproportionates above 273 K into CrO_2Cl_2 and a chromium(III) compound. CrOCl, and oxochloride of Cr^{III}, is obtained when Cr_2O_3 is heated strongly with $CrCl_3$.

OXIDES

All three metals form trioxides. White MoO_3 and yellow WO_3 are produced when the metals, their other oxides or their sulphides are heated in air or oxygen. The dark-red, crystalline CrO_3 is less thermally stable and cannot be made by direct combination of the elements; it is usually obtained by treating a saturated aqueous solution of a dichromate with concentrated H_2SO_4 :

$$Cr_2O_7^{2-} + 3\,H_2SO_4 \rightarrow 3\,HSO_4^- + H_3O^+ + 2\,CrO_3$$

It differs from MoO_3 and WO_3 in being very soluble in water, but it does not appear to form any crystalline hydrates. It is widely used in organic chemistry as a strong oxidising agent, usually in the form of a solution in acetic acid. The structure consists of chains of CrO_4 tetrahedra linked at corner oxygens; not surprisingly the *m.p.* is low (470 K). MoO_3 has a layer lattice and melts at 1070 K but WO_3 has an only slightly distorted version of the ReO_3 lattice and melts at 1450 K.

Hydrates of these oxides, the yellow $MoO_3 \cdot 2\,H_2O$ and the colourless, isomorphous $WO_3 \cdot 2\,H_2O$, can be crystallised from cold solutions of molybdates and tungstates when they are made strongly acidic. The molybdenum compound has been shown to contain sheets of MoO_6 octahedra sharing corners, with one H_2O covalently bound to Mo and the other hydrogen-bonded in the lattice. From hot solutions of molybdates and tungstates monohydrates $MO_3 \cdot H_2O$ are precipitated rapidly by acids.

When CrO_3 is melted it loses oxygen to give a series of lower oxides, including the ferromagnetic CrO_2 used in magnetic recording tapes, which has an undistorted routine structure, and eventually the green, thermally stable α-Cr_2O_3 which has the corundum lattice. This compound is more conveniently made by the action of heat on $(NH_4)_2Cr_2O_7$:

$$(NH_4)_2Cr_2O_7 \rightarrow N_2 + 4\,H_2O + Cr_2O_3$$

The oxides MoO_3 and WO_3 are reduced by heating with mild reducing agents. The Products are of great structural interest and have been extensively studied. For example, the yellow WO_3 changes to deep blue as a phase $W_{50}O_{148}$ crystallises. Fig. 4.4 illustrates the type of crystallographic shearing which is responsible for this phenomenon. As oxygen is lost from the trioxide, oxygen vacancies are removed by slabs of corner-linked octahedra moving into positions where they share edges with the octahedra of adjacent slabs. It has been shown that shearing of this type occurs at regular intervals throughout the solid phase ; the reasons for this long-range ordering of shear lines is not yet understood. Further reduction of both MoO_3 and WO_3 gives rise to phases of general formula M_nO_{3n-1} with $n = 14, 12, 11, 10, 9, 8$, and in the case of molybdenum $n = 5$ and 4 also. These last have more complicated structures in which the shearing is of a different kind, but long-range ordering again appears within a particular phase.

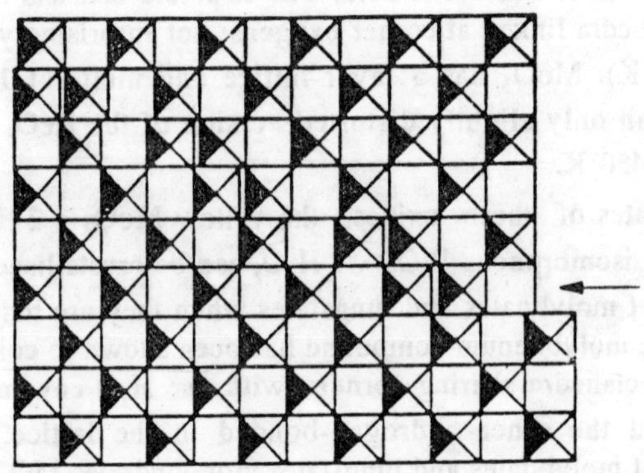

Fig. 4.4 Diagram indicating shearing in MO_3 -type oxide consisting of MO_6 octahedra sharing corners.

Further reduction with hydrogen at temperatures below 750 K, or with ammonia, produced the violet MoO_2 and brown WO_2. These compounds have much distorted routine structures in which the short distances between metal atoms indicates bonding between them.

CHROMATES, MOLYBDATES AND TUNGSTATES

Yellow solutions of the tetrahedral CrO_4^{2-} ion are turned orange below pH 6 because the dichromate ion, $Cr_2O_7^{2-}$ is produced :

$$2\,CrO_4^{2-} + 2\,H^+_{aq} \rightleftharpoons Cr_2O_7^{2-} + aq$$

The equilibrium is labile, however, and insoluble chromates like Ag_2CrO_4, $BaCrO_4$ and $PbCrO_4$ are precipitated even from mildly acidic solution. The last of these is known as chrome yellow, a useful pigment. A salt $KCrO_3Cl$ can be crystallised from a solution of $K_2Cr_2O_7$ in strong aqueous HCL. Despite the fact that $E°$, $Cr_2O_7^{2-}/Cr^{3+}$ is higher than $E°$, X_2/X^- for X = Cl, Br or I, all three compounds $KCrO_3X$ can be made similarly because the redox reactions are so slow.

Normal molybdates and tungstates containing discrete, tetrahedral MoO_4^{2-} and WO_4^{2-} ions can be crystallised from solutions with limited pH ranges. The salt $(NH_4)_2MoO_4$ separates from a solution of MoO_3 in strong aqueous ammonia, but from a near-neutral solution the salt which is obtained is $(NH_4)_6Mo_7O_{24} \cdot 4\,H_2O$. The $Mo_7O_{24}^{6-}$ ion is an example of an isopolyanion. Octamolybdates containing $Mo_8O_{26}^{4-}$ can also be made, and the most important species obtained in tungstate solutions are $HW_6O_{21}^{5-}$ and $W_{12}O_{41}^{10-}$. These ions are probably hydrated to some extent both in solution and in crystals. It has been shown, for example, that the ion $H_2W_{12}O_{42}^{10-}$ is present in the salt usually formulated $(NH_4)_{10}W_{12}O_{41} \cdot 11\,H_2O$.

HETEROPOLYACIDS

The yellow precipitate obtained in the ammonium molybdate test for a phosphate is $(NH_4)_2H(PMo_{12}O_{40}) \cdot H_2O$. When it is washed with dilute NH_4NO_3 it becomes $(NH_4)_3(PMo_{12}O_{40})$; it is in fact a useful inorganic cation exchanger. This 12-molybdophgosphate is an example of a large number of heteropolysalts formed by molybdates and tungstates when acidified in the presence of other oxo-anions. Some typical heteropolyacids are

$H_4(SiMo_{12}O_{40})$ 12-molybdosilicic acid

$H_3(AsMo_{12}O_{40})$ 12-molybdoarsenic acid

$H_5(BW_{12}O_{40})$ 12-tungstoboric acid

They are known as 12-acids because there are twelve Mo or W atoms to one hetero-atom. More than thirty elements are known to function as heteroatoms in these compounds ; examples are Ti, Ge,

Sn, Zr, Hf in their +4 states. The most common structure is shown in Fig. 4.5. It consists, in effect, of a tetrahedron around the hetero-atom, every corner of which is shared by three WO_6 or MoO_6 octahedra each of which shares an oxygen with its neighbours. The four resulting Mo_3O_{13} groups, for example, share corners to form $(MMo_{12}O_{40})^{n-8}$ ions where *n* is the charge number of M. The large open spaces in ions of this type allow the inclusion of water molecules and cations. Hydrates are, in fact, common and the insoluble salts are often excellent cation-exchangers. There are a few 12-acids containing the ions $(MW_{12}O_{42})^{n-12}$ and $(MMo_{12}O_{42})^{n-12}$ where M is, for example, Ce^{IV} or Th^{IV}, in which the heteroatoms appear to be octahedrally co-ordinated to oxygen atoms instead of tetrahedrally.

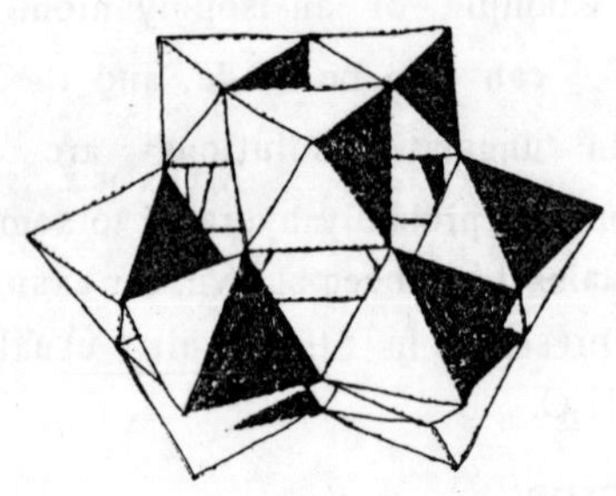

Fig. 4.5. Most common structure of a 12-acid heteropolyanion.

6-acids of molybdenum are also known; they have the general formula $H_m(MMo_6O_{24})$, where M can be I, Te, Fe, Co, Al, Rh, Cr, Cu or Mn, and its charge number 12—*m*. In the 6-acids, six MoO_6 octahedra are joined by sharing edges to form a hexagonal annulus which provides 6 oxygen atoms to coordinate the central hetero-atom (Fig. 4.6). The structure of the isopolyanion $Mo_7O_{24}{}^{6-}$ is just this with an Mo atom replacing the hetero-atom at the centre.

Molybdenum also forms salts of 9-acids, e.g. $H_6(MnM\ {}_9O_{32})$ and salts of the more complex 10- and 11-acids with Mo : hetero-atom ratios of 10 and 11 respectively. Dimeric ions derived from the 9-acids, e.g. $(P_2Mo_{18}O_{62})^{6-}$, have also been studied.

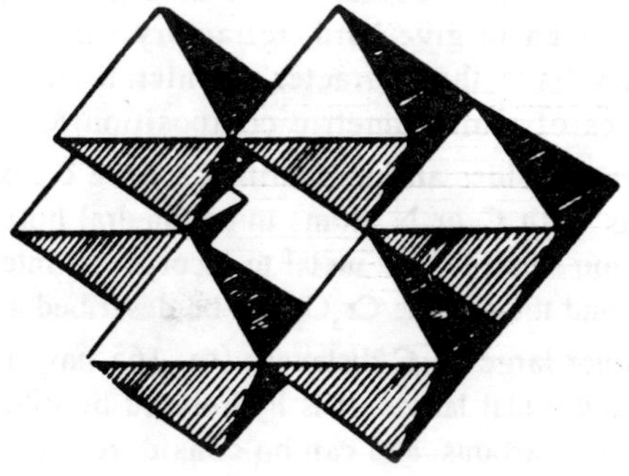

Fig. 4.6. Arrangement of MoO_6 octahedra in a 6-acid.

TUNGSTEN BRONZES

When alkali-metal tungstates are reduced with hydrogen at red heat, or when WO_3 is treated with the vapour of the alkali metal, intensely coloured, unreactive substances of bronze-like lustre are obtained. These tungsten bronzes have the general formula M_xWO_3 (M = Li, Na or K and $x < 1$) and are of considerable interest structurally. When $x > 0.3$ the lattice consists of cubic ReO_3 unit cells with a fraction x of them containing alkali metal atoms at their body centres. These atoms evidently contribute their 3s electrons to delocalised energy bands rather like those in a metal, and the compounds are good electrical conductors. For values of x below 0.3 the cubic lattice is distorted towards triclinic, as in WO_3 itself, the conductivity falls, and the colour tends towards violet. It has not been found possible to make compounds with $x < 0.95$. These compounds rich in alkali metal golden-yellow ; this structure approximates closely to that of perovskite and the cell dimensions are greatest when x is greatest.

BINARY COMPOUNDS WITH OTHER NON-METALS

Of the Group VI metals, only chromium forms a hydride, which can be made by electrolytic reduction of certain Cr^{VI} solutions. A phase with composition range $CrH_{0.5-1.0}$ has a hexagonal structure; a second less thermally stable phase with a composition $CrH_{1.0}$ to $CrH_{1.7}$ has an anion-deficient modification of the fluorite structure.

Molybdenum and tungsten react at high temperatures with carbon and nitrogen to give hard, refractory carbides and nitrides. These generally have the character of interstitial compounds, but there are phases of stoichiometric composition M_2C, M_2N,MC.and MN These monocarbides and mononitrides have c.c.p. arrangements of metal atoms with C or N atoms in octahedral holes. The metallic radius of Cr is too small for the metal to incorporate interstitial C atoms in its structure, and the carbide Cr_3C_2 can be described as having carbon chains, with rather large C- C distances (ca. 165 pm), running through a grealty distorted metal lattice. It is hydrolysed by dilute acids to give a mixture of hydrocarbons, and can be considered to be intermediate between the truly ionic carbides and the interstitial ones.

There are several types of interstitial borides. MoB and WB have structures like FeB whereas MoB_2 and WB_2 resemble AlB_2.

The chromium—sulphur system is complex. Dark-green, paramagnetic Cr_2S_3 can be made by the action of H_2S on Cr_2O_3. When it is heated to a high temperature it can be converted eventually into CrS, but there is evidence for several intermediate phases forming a homologous series Cr_nS_{n+1}.

The most important sulphide of molybdenum is molybdenite, MoS_2, which has a structure rather like that of CdI_2, but with the layers of S atoms eclipsed instead of staggered (Fig. 4.7). Molybdenite resembles graphite in being an excellent lubricant, and is added to engine oils for that purpose. A trisulphide, brown MoS_3, is precipitated when H_2S is passed into acidified aqueous molybdate. The sulphides WS_2 and WS_3 resemble the molybdenum compounds.

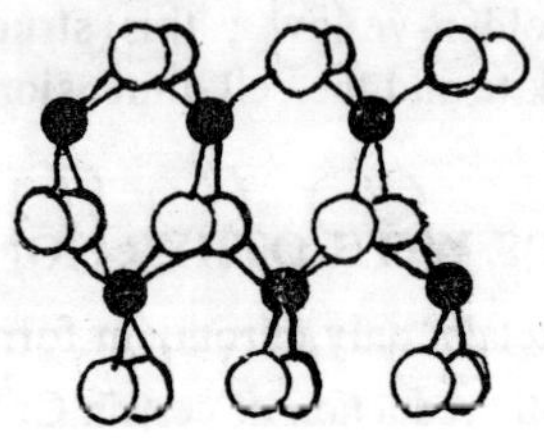

Fig. 4.7. Eclipsed arrangement of S atoms in molybdenite, MoS_2.

ORGANOMETALLIC COMPOUNDS AND π - COMPLEXES

Dibenzenechromium, which has a sandwich structure with a Cr atom between parallel benzene ring, can be made by heating $CrCl_3$ with aluminium, Al_2Cl_6 and benzene to obtain $[(\pi-C_6H_6)_2Cr]^+$ $(AlCl_4)^-$ and then reducing this with aqueous sodium dithionite. The dark-brown, diamagnetic solid can be sublimed in vacuo at 420 K but is much more sensitive to air than is ferrocene. Although $(C_6H_6)_2Cr$ contains the same number of bounding electrons as ferrocene, it is much less stable to electrophiles ; attempts to effect aromatic substitution always cause decomposition. Bisbenzenoid complexes of Mo and W have also been made ; they too are unstable towards electrophiles.

Sodium cyclopentadienide reacts with anhydrous $CrCl_2$ in tetrahydrofuran to given scarlet $(\pi-C_5H_5)_2Cr$. Compounds containing a blue cation, probably $[(\pi-C_5H_5)CrCl(H_2O)_n]^+$, are easily obtained on oxidation in the presence of HCl. The compound $(\pi-C_5H_5)_2Cr$ is, in fact, very air-sensitive. The metals of this group form many π -cyclopentadienyl carbonyl complexes. The infrared spectrum of $[(\pi-C_5H_5)Cr(CO)_3]_2$ shows no bridging carbonyl frequency, and the analogous Mo compound has been shown by X-ray analysis to have an Mo—Mo bound. The group $(\pi-C_5H_5)Mo(CO)_3$ is surprisingly inert ; the dimer can be converted to $Na[(\pi-C_5H_5)Mo(CO)_3]$ by sodium amalgam in tetrahydrofuran and to $(\pi-C_5H_5)Mo(CO)_3H$ by hydrogen at 17 MPa. This hydride can be converted to the bromo-derivative $(\pi-C_5H_5)Mo(CO)_3Br$ with *N*-bromosuccinimide and into $(\pi C_5H_5)Mo(CO)_3CF_2CF_2H$ with tetrafluoroethylene. The compound $(\pi-C_5H_5)_2MoH_2$, obtained from $MoCl_5$, C_5H_5Na and $NaBH_4$ in tetrahydrofuran, has the structure shown in Fig. 4.8. The two C_5H_5 rings are inclined at 34° to one another.

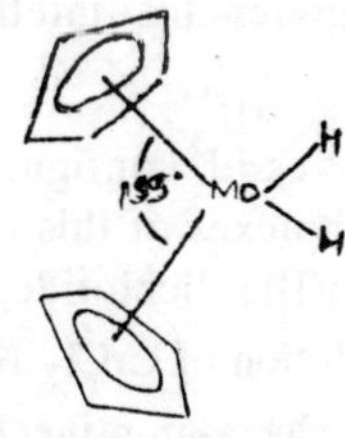

Fig. 4.8. Structure of $(\pi-C_5H_5)_2MoH_2$.

Chromium and molybdenum compounds in which the metals are π-bonded to seven-membered and eight-membered cyclic hydrocarbons have also been made. Thus (π $-C_5H_5$)(π $-C_7H_7$)Cr has been made by the method which is out lined :

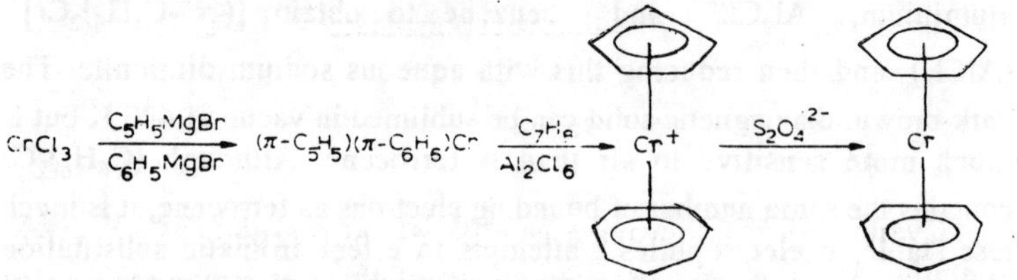

$$CrCl_3 \xrightarrow[C_6H_5MgBr]{C_5H_5MgBr} (\pi-C_5H_5)(\pi-C_6H_6)Cr \xrightarrow[Al_2Cl_6]{C_7H_8} Cr^+ \xrightarrow{S_2O_4^{2-}} Cr$$

The compound $C_8H_8Mo(CO)_3$ has been shown to have the structure shown in Fig. 4.9, in which the cyclo-octatetrene acts as a 6-electron donor.

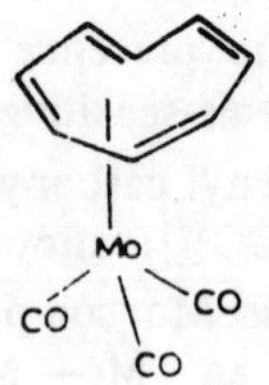

Fig. 4.9. Structure of $C_8H_8Mo(CO)_3$.

COMPLEXES

There are not many complexes containing the metals in the + 2 state. Chromium forms ammines, for instance $Cr(NH_3)_6Cl_2$, which are high-spin d^4 complexes and, as such, exhibit Jahn—Teller distortion to a tetragonal symmetry. There are some low-spin cyanocomplexes such as $K_4Cr(CN)_6$. Molydenum(II) is represented by high-spin orthophenylenebisdimethyldiarsine complexes, $Mo(diars)_2X_2$ (X = Cl, Br, I).

Chromium(III) complexes (d^3 configuration) are very common ; there are many ammine complexes of this oxidation state. The purple chloropentamminechromium(III) dichloride, $[Cr(NH_3)_5Cl]Cl_2$, is made by bubbling air through a solution of $CrCl_2$ NH_4Cl and NH_3 in water. It can be converted into hexa-amminechromium(III) trichloride,

$[Cr(Nh_3)_6]Cl_3$ (yellow), by treating its cold, concentrated solution with ammonia. A violet dichlorotetra-amminechromium(III) chloride, $[Cr(NH_3)_4Cl_2]Cl$, exists and also the triammine, $Cr(NH_3)_3Cl_3$. Werner (1910) made the latter by the reactions :

$$CrO_3 \xrightarrow[\text{dilute } H_2SO_4]{H_2O_2 \text{ pyridine and}} \text{pyridinium perchromate} \xrightarrow{NH_3} CrO_4(NH_3)_3$$

$$\downarrow \text{cold concentrated HCl}$$

$$Cr(NH_3)_3Cl_3$$

Cyano- and thiocyanato-complexes of chromium are also common. Among the more interesting of these is Reinecke's salt, $NH_4[Cr(NH_3)_2(SCN)_4]H_2O$, made by adding $(NH_4)_2Cr_2O_7$ slowly to melted NH_4SCN, washing with, and recrystallising from, alcohol. The octahedral ion has the form shown in Fig. 4.10.

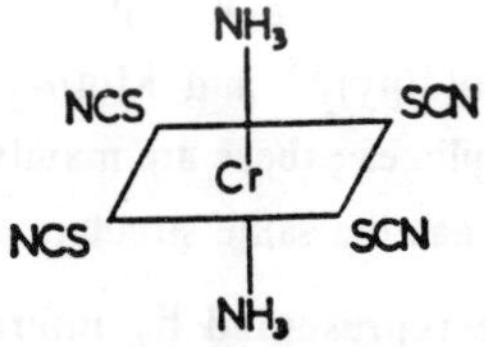

Fig. 4.10. Structure of $[Cr(NH_3)_2(SCN)_4]^-$ ion present in Reinecke's salt.

Some of the oxalato-complexes of chromium(III) are also of interest. Potassium trisoxalatochromate(III), $K_3Cr(C_2O_4)_3 \cdot 3H_2O$, is obtained by adding potassium oxalate to the solution obtained by reducing $K_2Cr_2O_7$ with oxalic acid :

$$K_2Cr_2O_7 + 7H_2C_2O_4 \rightarrow K_2C_2O_4 + Cr_2(C_2O_4)_3 + 6CO_2 + 7H_2O$$

$$\downarrow 2K_2C_2O_4$$

$$2K_3[Cr(C_2O_4)_3]$$

The anion of the blue crystalline compound was resolved by Werner (1912) into dextrorotatory and laevorotatory forms (Fig. 34. 11A). Potassium dioxalatodiaquochromate, $K[Cr(C_2O_4)_2(H_2O)_2]$, exists in *cis*- and *trans*-forms, the former (Fig. 4.11B (i)) showing purple-green dichroism and the latter (Fig. 4.11B (ii)) being mauve.

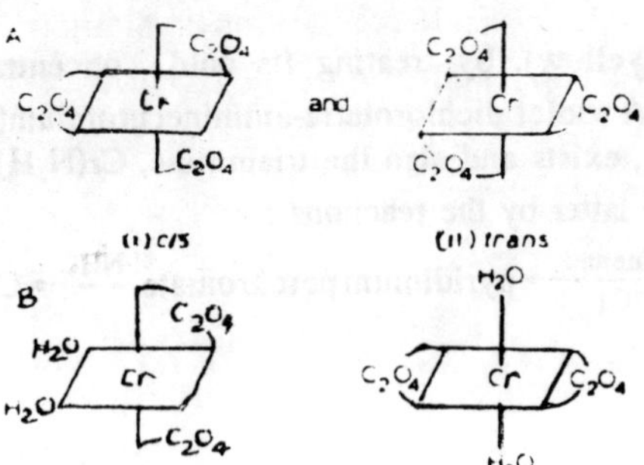

Fig. 4.11. A. *d*- and *l*-forms of $[Cr(C_2O_4)_3]^{3-}$ ion. B. *cis*- and *trans*-forms of $[Cr(C_2O_4)_2\text{-}(H_2O)_2]^-$ ion.

Molybdenum(III) exists in neutral, anionic and cationic complexes. $MoCl_3$ and $MoBr_3$ form the adducts with pyridine $Mo(py)_3X_3$, but the halides undergo solvolysis in liquid ammonia, methylamine and dimethylamine to give products such as $MoBr_2NH_2$, $MoBr_2NHMe$ and. $MoBr_2NMe_2$ The complex anions include many octahedral ones of the MoX_6^{3-} type in which X is a halogen or pseudohalogen. The complex Mo^{III} cations include $Mo(dipy)_3^{3+}$ and $Mo(o\text{-phen})_3^{3+}$. Tungsten(III), however, forms few complexes ; these are mainly complex halides. The $W_2Cl_9^{3-}$ ion in $K_3W_2Cl_9$ has the same structure as the $Tl_2Cl_9^{3-}$ ion.

Mo^{IV} and W^{IV} are represented by interesting 8-co-ordinated cyanocomplexes containing $M(CN)_8^{4-}$ ions. $K_4Mo(CN)_8$ is made by treating a K_3MoCl_6 solution with KCN in the presence of air. These octacyanocomplexes are remarkably thermally and hydrolytically stable. The free acid $H_4[Mo(CN)_8] \cdot 6\,H_2O$, which can be isolated as crystals when the potassium salt is acidified with HCl, can be oxidised to $H_3[Mo(CN)_8] \cdot 3\,H_2O$ by strong oxidising agents such as MnO_4^-. The Mo^{5+} ion has only one 4d electron, and the dodecahedral arrangement of CN^- ligands around it should stabilise one 4d orbital, capable of accommodating these electrons, relative to the other four :

Energy ↑

d_{xy}, d_{yz} — —

d_{z^2} —

$d_{x^2-y^2}$ —

d_{xz} —

The structure of the $Re(CN)_8^{3-}$ ion is similarly stabilised. In solid $K_4Mo(CN)_8$ the complex ion is dodecahedral (Fig. 4.12).

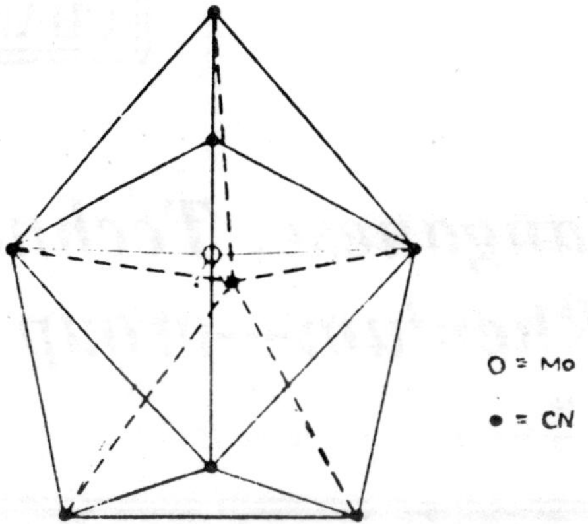

Fig. 4.12. Dodecahedral structure of $Mo(CN)_8^{4-}$ ion.

The oxidation state +5 is represented by some fluoro- and oxohalide complexes of molybdenum and tungsten, such as M^IMoF_6, M^IWF_6, $M^I_2[MoOCl_5]$ and $M^I[MoOBr_4]$. There are also the cyanocomplexes like $K_3W(CN)_8$, made by the oxidation of aqueous solutions of the corresponding +4 complexes (above).

Molybdenum(VI) and tungsten(VI) exist in anionic complexes containing oxygen and fluorine as ligands. Examples are $[MoOF_5]^-$ and $[WOF_5]^-$, $[MoO_3F_3]^{3-}$ and $[WO_3F_3]^{3-}$.

CHAPTER 5

Manganese, Technetium and Rhenium—group VIIA

INTRODUCTION

As in the preceding groups the properties of the first member stand apart from those of the higher congeners. All three elements have seven d and s electrons available for valency orbitals and the +7 state is clearly characterised in the MnO_4^-, TeO_4^- and ReO_4^- ions. Tc and Re differ from Mn in failing to form simple +2 ions. In this respect they resemble the elements of corresponding series of the earlier transition-metal groups.

TABLE 5.1 : ATOMIC PROPERTIES OF Mn, Te AND Re

	Mn	Tc	Re
Z	25	43	75
Electron configuration	[Ar] $3d^54s^2$	[Kr]$4d^55s^2$	[Xe] $4f^{14}5d^56s^2$
$I(1)$kJ mol^{-1}	717	703	760
Metallic radius/pm	126	136	137

Physically, the metals are typical d-block metals; their densities (Table 5.2) are normal for the positions they occupy in the Periodic Table. The *m.p.* of Tc and Re are only slightly less than those of Mo

TABLE 5.2 : PHYSICAL PROPERTIES OF Mn, Tc and Re

	Mn	Tc	Re
ρ/g cm^{-3}	7.21–7.44	11.5	21.0
M.p./K	1517	2600	3450

and W which precede them, but that of manganese is low compared with the metals on either side of it. Cr and Fe. Although technetium and rhenium have h.c.p. structures, manganese exists in three forms, not one of which has the simple 8-co-ordination or 12-co-ordination typical of the transition metals. In α-Mn, for example there are four kinds of crystallographically non-equivalent atoms and the Mn–Mn distances vary from 224 pm to 296 pm. Its brittleness and its resistance to wear may both be due to this peculiarity.

In Group VIIA the first ionisation energy of the second element, Tc, is rather less than that of the first element, Mn. In this respect the group is different from Group VIA but similar to the iron and cobalt groups. The effect on the chemistry is minimal, however. The value of $E^\circ, Mn^{2+}/Mn$ (Table 5.3) indicates manganese to be a better reducing agent than its neighbours, a fact which must be largely due to the lower boiling point and correspondingly low sublimation energy of the metal.

TABLE 5.3 : IONISATION ENERGIES AND ELECTRODE POTENTIALS OF GROUP VIIA ELEMENTS

		Mn	Tc	Re
Ionisation energy	$I(1)$/kJ mol^{1}	717	703	760
	$I(2)$/kJ mol^{-1}	1492	1470	
E°, M^{2+}/M/V		–1.18		
E°, MO_2/M/V			+0.27	+0.25
E°, MO_4/M/V		+0.79	+0.47	+0.34

The redox potentials for technetium and rhenium (Table 5.3) are for the reactions :

$$MO_2 + 4H_3O^+ + 4\,e = M + 6\,H_2O$$

$$MO_4^- + 8\,H_3O^+ 7\,e = M + 12 + H_2O$$

The free energies of some of the oxidation states relative to the metal in aqueous solution at pH 0 are given in Fig. 5.1. A striking point is the very great stability of Mn^{II} to both oxidation and reduction. The Mn^{VI} state is unstable to disproportionation in acid solution but is more stable at high pH. Thus K_2MnO_4 (which is isomorphous with K_2CrO_4) is converted by as weak an acid as H_2CO_3 into $KMnO_4$ and MnO_2.

$$3\,K_2MnO_4 + 2\,CO_2 \rightarrow 2\,KMnO_4 + MnO_2 + 2K_2CO_3$$

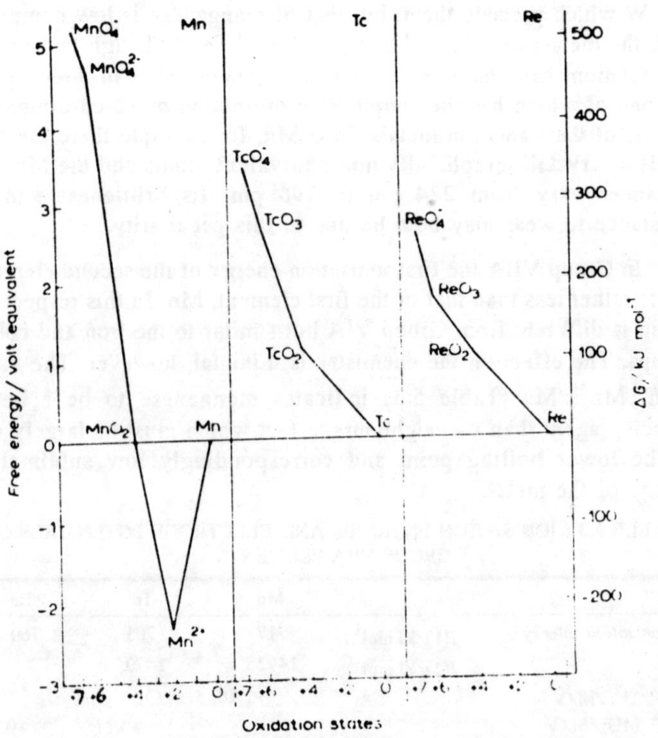

Fig. 5.1. Free energies of oxidation states relative to the metal in aqueous solution at pH = 0.

Technetium and rhenium salts corresponding to Mn^{2+} salts are not known. The formal charge +7 is dominant in Tc and Re; compounds in this state have far less oxidising power than MnO_4^-. Free permanganic acid exists only in aqueous solution and the oxide decomposes explosively above 273 K. The corresponding compounds of Tc and Re are, however, stable.

THE ELEMENTS

Manganese

The element (0.085% of the lithosphere) is the most common transition metal after iron and titanium. Four of its five oxides exist in nature but the most important ore is pyrolusite, a tetragonal form of MnO_2. The principal metallurgical form of manganese is ferromanganese (ca. 80% Mn) which is made by reducing MnO_2 and Fe_2O_3 in a furnace with coke in the presence of dolomite which removes the SiO_2 as a slag. Spiegeleisen (5–20% Mn, 3–5% C) is made by a modification of the process. Purer manganese (99.9% Mn) is manufactured when required by the electrolysis of aqueous $MnSO_4$. Almost every grade of steel contains manganese. It combines with sulphur which would otherwise remain as FeS and make the steel brittle when hot; the MnS forms harmless inclusions, thereby improving the rolling and forging properties of the alloy. Furthermore, Mn acts as a deoxidiser when the metal is molten and it also improves the strength, toughness and response to heat treatment after solidification.

The metal is reactive ; it combines with the halogens, oxygen, sulphur, carbon, nitrogen and many metalloids. It liberates H_2 from cold dilute HCl and H_2SO_4 and from steam at red heat.

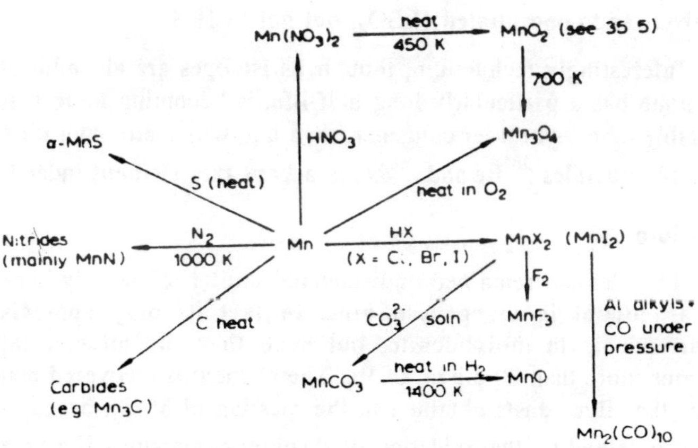

Fig. 5.2. An outline of manganese chemistry.

Technetium

The element occurs in trace amounts of earth as Tc-99, a fission product of uranium, and has also been observed spectroscopically in the sun and some stars. In 1939 the 90-days ^{97m}Tc was made by bombarding molybdenum with high-energy deuterons from a cyclotron. The long-lived ^{99}Tc ($t_{1/2}$ = 2.12×10^5 y) forms about 6% of the fission products of uranium, and a reactor operating at 100 MW yields about 2.5 g per day. This separated from the uranium and its fission products, is the principal source of the metal.

The hot, acidic solution of fission products is treated with tetraphenyl-arsonium chloride in the presence of an excess of $HClO_4$, and a precipitate of Ph_4AsTcO_4 in carrier Ph_4AsClO_4 obtained, leaving most of the other fission products (*e.g.* Cs, Sr and lanthanides) in solution. The Tc is recovered by dissolving the precipitate in H_2SO_4 and electrolysing between Pt electrodes. The black TcO_2 which is deposited is converted to Tc_2O_7 by dissolving in $HClO_4$, and separated by precipitating as Tc_2S_7 with H_2S.

$$2\,HTcO_4 + 7\,H_2S \rightarrow Tc_2\,S_7 + 8\,H_2O$$

The metal can be made by reducing Tc_2S_7 with hydrogen at 1400 K. It is bright and silvery, tranishes in moist air, burns in O_2 to give Tc_2O_7, in fluorine to give TcF_5 and TcF_6, and combines with S to give TcS_2 and with C to give TcC at high temperatures. It dissolves in HNO_3 and concentrated H_2SO_4, but not in HCl.

Interestingly technetium, though its isotopes are all radioactive and none has a particularly long half-life, is becoming more readily available than its heavier congener, rhenium, which although it exists as stable nuclides ^{185}Re and ^{187}Re, is a very rare element indeed.

Rhenium

The element remained undiscovered until 1925 largely because it was sought in manganese ores. In fact its only appreciable occurrence is in molybdenite, but even then the mineral rarely contains more than 20 p.p.m. of Re. The element is recovered mainly from the flue dusts obtained in the roasting of MoS_2, because the Re_2O_7 obtained by the oxidation of rhenium compounds is a volatile substance. The metal is usually made by reducing NH_4ReO_4 with hydrogen at 700 K. Like Tc it dissolves in HNO_3 and H_2SO_4, but not in HCl. Unlike Tc it dissolves easily in H_2O_2, the product being $HReO_4$, but in its other reactions it resembles Tc closely Rhenium is a very expensive metal and has found little commercial use, though it may have value as a catalyst for hydrogenation and dehydrogenation reactions.

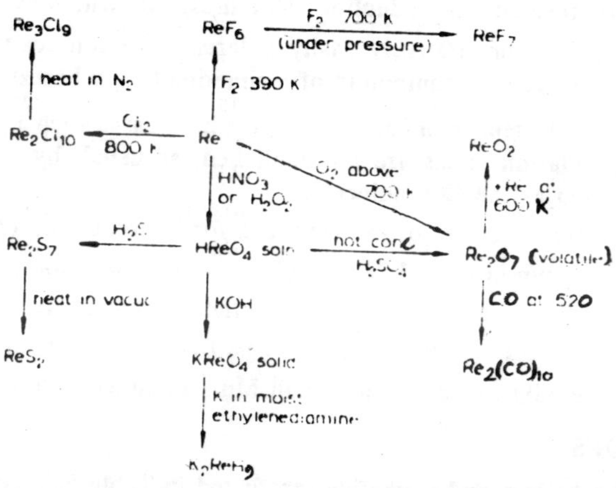

Fig. 5.3 : An outline of rhenium chemistry.

TABLE 5.4 COMPOUNDS AND IONS REPRESENTING OXIDATION STATES OF Mn, Tc AND Re

Charge number	Mn	Tc	Re
—1	$Mn(CO)_5^-$		$Re(CO)_5^-$
0	$Mn_2(CO)_{10}$	$Tc_2(CO)_{10}$	$Re_2(CO)_{10}$
+ 1	$Mn(CO)_5Cl$	$K_5Tc(CN)_6$	$Re(CO)_5Cl$
+ 2.	$Mn(H_2O)_6^{2+}$	$(\pi\ C_5H_5)_2Tc$	$ReCl_2$ (diars)
+ 3	MnF_3	$Tc(diars)_2Cl_2^+$	Re_3Cl_9
+ 4	MnO_2	$TcCl_4$	K_2ReCl_6
+ 5	Na_3MnO_4	$TcOBr_3$	$ReOCl_4^-$
+ 6	MnO_4^{2-}	TcF_6	ReO_3
+ 7	MnO_4^-	Tc_2O_7	ReO_3Cl

OXIDATION STATES

The representative compounds and ions shown in Table 5.4 illustrate the usual trends shown by transition metals near the middle of the d block. The highest possible oxidation state, numerically equal to the group number, is shown by all three elements. There are several indications in the table that the higher oxidation states of Tc and Re are more resistant to reduction than those of Mn, however. For example, the comparatively easily polarised Br^- ion features as a ligand in $TcOBr_3$, a compound of technetium(V), whereas the only ligands which stabilise manganese(V) are hard bases such as O^{2-}. The lowest oxidation states are all stablished, as usual, by π-acceptor ligands such as the CO molecule.

The free energies of the reactions for the reduction of H^+aq at pH 0 are shown in Fig 5.1. Tc^{VII} and Re^{VII} are clearly not nearly such strong oxidising agents as MnO_4^- ; another feature of the diagram is that it shows the +6 state in Tc, and particularly in Re, to be much more thermodynamically stable than Mn^{VI} in aqueous solution.

HALIDES

The halides and oxohalides are listed in Table 5.5. Features of note are (a) the absence of higher iodides, (b) the absence of high oxidation states of Mn except in MnO_3F, (c) the existence of many oxofluorides of rhenium.

TABLE 5.5 : HALIDES AND OXOHALIDES OF MANGANESE, TECHNETIUM AND RHENIUM

Charge number	Mn	Tc	Re
+2	MnF_2, $MnCl_2$, $MnBr_2$, MnI_2		
+3	MnF_3		Re_3Cl_9, Re_3Br_9, Re_3I_9
+4	MnF_4	$TcCl_4$	ReF_4, $ReCl_4$, $ReBr_4$, ReI_4
+5		TcF_5	ReF_5, $ReCl_5$, $ReBr_5$
		$TcOCl_3$, $TcOBr_3$	$ReOF_3$
+6		TcF_6	ReF_6, $ReCl_6$
		$TcOF_4$	$ReOF_4$, $ReOCl_4$, $ReOBr_4$
+7			ReF_7
			$ReOF_5$
			ReO_2F_3
	MnO_3F	TcO_3F, TcO_3Cl	ReO_3F, ReO_3Cl, ReO_3Br

The dihalides of Mn, except MnF_2 which is insoluble and is usually made by treating $MnCO_3$ with HF, are obtained as rose-pink hydrates when the metal is dissolved in the appropriate aqueous hydrogen halide.

The red-purple MnF_3, the only trihalide of Mn, is made by the action of F_2 on $MnCl_2$ or MnI_2. With water it gives MnO_2 and MnF_2. The hygroscopic, blue MnF_4 may be made by fluorinating powdered MnF_3 at 800 K. At room temperature it slowly decomposes to MnF_3 and fluorine.

The rhenium(III) halides are of some interest structurally Re_3Cl_9, made by heating $ReCl_5$ in nitrogen, is a dark-red solid containing the trimeric units shown in Fig. 5.4a. Half of the terminal Cl atoms are used in bridging to adjacent clusters (Fig. 5.4b) ; Re_3Br_9 is similar, but Re_3I_9, made by the decomposition of ReI_4 in a sealed tube at 600 K, is rather different, only two rhenium atoms in each cluster being chlorine-bridged to neighbouring clusters.

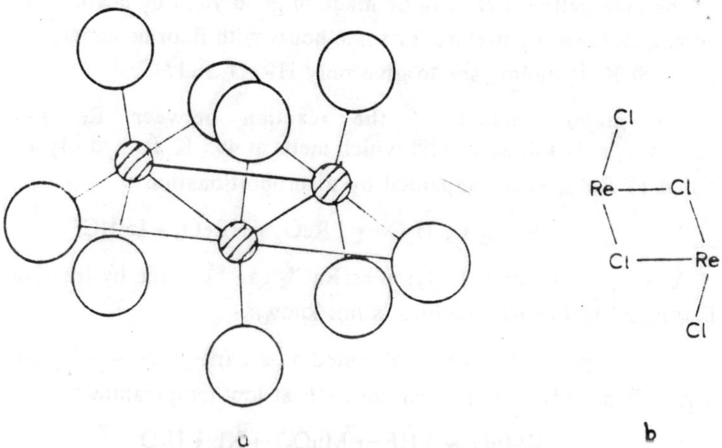

Fig. 5.4. Re_3Cl_9. (a) Trimeric unit in the solid. (b) Showing how terminal Cl atoms are used as bridges to adjacent clusters.

$TcCl_4$ is the only thermally stable chloride of technetium yet made. The red crystals can be obtained by heating Tc_2O_7 to 700 K with CCl_4 in a sealed tube. The slightly distorted $TcCl_6$ octahedra which constitute the structure are joined into a chain by sharing edges. Of the tetrahalides of Re, the bromide and iodide can be made by treating $HReO_4$ with the respective hydrogen halide, and ReF_4 by reducing ReF_6 with Re. The amorphous, black compound known as α–Re Cl_4 is made by treating ReO_2 with $SOCl_2$. A crystalline material known as β – $ReCl_4$, which may be the same compound, has been found to contain Re_2Cl_8 units linked into chains by single chlorine bridges.

When technetium powder reacts with fluorine the major product is the golden-yellow solid TcF_6 and the minor product the yellow solid TcF_5. There is no evidence for a heptafluoride, Rhenium, however, gives a mixture of ReF_6 and ReF_7. The former, a low-melting yellow solid, is very reactive ; it hydrolyses in water :

$$3\,ReF_4 + 10\,H_2O \rightarrow 2\,HReO_4 + ReO_2 + 18\,HF$$

and reacts with silica :

$$2\,ReF_6 + SiO_2 \rightarrow 2\,ReOF_4 + SiF_4$$

The pale-yellow ReF_7 can be made in good yield by heating the foregoing ReF_6/ReF_7 mixture for some hours with fluorine under 300 kPa at 670 K. It hydrolyses to give only $HReO_4$ and HF.

The major product of the reaction between Re and Cl_2 is $ReCl_5$, a dark-brown solid which melts at 495 K. Its hydrolysis, like that of ReF_6, is accompanied by disproportionation :

$$3\,ReCl_5 + 8\,H_2O \rightarrow 2\,ReO_2 + HReO_4 + 15\,HCl$$

A compound which analyses as $ReCl_6$ can be made by treating ReF_6 with BCl_3, but its structure is not known.

The compound MnO_3F is obtained as a dark-green solid (*m.p.* 195 K) when $KMnO_4$ is treated with HF at low temperature :

$$KMnO_4 + 2\,HF \rightarrow MnO_3F + KF + H_2O$$

It decomposes explosively at room temperature. A compound claimed to be MnO_3Cl has been made by treating Mn_2O_7 with $ClSO_2OH$. The oxofluorides of Tc and Re are more thermally stable TcO_3F, a yellow solid, is obtained when TcO_2 is heated with F_2. The colourless liquid ReO_3Cl, made by reaction between Re_2O7 and $ReCl_5$, can be converted to the yellow solid ReO_3F by treatment with

HF. A minor product of the reaction is $ReOF_5$. The compound $ReOF_4$ is well characterised ; its unit of structure is a square pyramid :

The elements display a wide range of complex halides. Manganese(II) forms complexes of the types M^1MnF_3, $M^1_2MnCl_4$ and $M^1_4MnCl_4$. Moreover, $MnCl_2$ and $MnBr_2$ react with salts like the pyridinium and tetramethylammonium halides to give tetrahedral MnX_4^{2-} complexes, Manganese (III) complexes such as $M^1_2MnF_5$ and $M^1_2MnCl_5$ are known. When MnO_2 is added to a solution of HCl in CCl_4, a dark-green colour is produced and the salt $(Et_4N)_2MnCl_5$ can be extracted after addition of Et_4NCl.

Rhenium in its +2 and +3 states forms a number of complex halide ions in which there are strong metal–metal interactions. The ions $Re_2Cl_8^{2-}$ and $Re_3Cl_9^{3-}$ are discussed in Paragraph 27.12. The ion in the compound $CsReCl_4$ is a trimeric one, $Re_3Cl_{12}^{3-}$, in which there is a triangle of Re atoms each of which is seven-co-ordinate (Fig. 5.5).

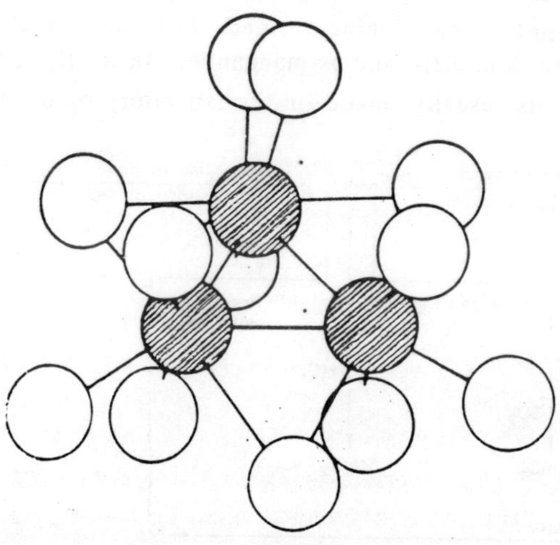

Fig. 5.5 The $Re_3Cl_{12}^{3-}$ anion in $CsReCl_4$.

The most common complex halides of Mn and Re are those in which the metal has a +4 charge ; examples are $M^{1}_{2}MnCl_6$, $M^{1}_{2}ReCl_6$ and, particularly, the corresponding fluoro- complexes. This +4 state of Re is obtained from states of higher charge in the formation of these compounds :

$$2\,ReCl_5 + 4\,KCl \rightarrow 2\,K_2\,ReCl_6 + Cl_2$$

The fluoro-compound K_2ReF_6 is made in good yield by the action of anhydrous HF of K_2ReI_6, but the acid $H_2\,Re\,F_6$ cannot be isolated.

The Mn^{IV} complexes are spin-free with 3 unpaired electrons. The magnetic susceptibilities of the complex chlorides of Re^{IV} and Tc^{IV} increase with temperature and the magnetic moment approaches that expected for 3 unpaired electrons. There is evidently some slight metal–metal interaction even in the +4 state.

OXIDES

Manganese occurs as manganosite, MnO, with the NaCl structure, hausmannite, Mn_3O_4, with a normal spinel structure which is distorted towards the tetragonal, braunite, Mn_2O_3, with the structure of a C-type lanthanide oxide and pyrolusite, MnO_2, which has a distorted rutile lattice. Gross nonstoichiometry is common, for example the composition of MnO can be increased to $MnO_{1.13}$ without a new phase being formed. The element also occurs as pyrochroite, $Mn(OH)_2$, and as manganite, MnO(OH). The dull-green monoxide is usually made in the laboratory by decomposing the

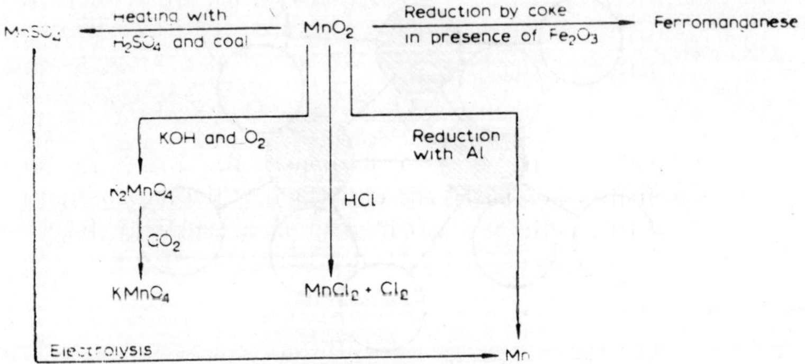

Fig 5.6. Reactions of manganese dioxide

carbonate in a nitrogen atmosphere. The corresponding hydroxide, precipitated by OH^- from Mn^{2+} solutions, rapidly darkens in colour in air to give MnO(OH) which can be converted at 500 K into Mn_2O_3, which is believed to contain Mn^{II} and Mn^{IV} but not Mn^{III}. Black Mn_3O_4 is made by heating any of the other oxides to 1300 K. The principal one, pyrolusite, is used (a) for making ferromanganese and manganese itself, (b) as a dipolariser in dry cells, (c) for rendering glass colourless, (d) as a drier in paint and (e) for making $MnSO_4$ which is used in treating manganese-deficient soils. Pune MnO_2 is made by decomposing $Mn(NO_3)_2$ at 450 K and then dehydrating in oxygen at 750 K, but the product is always stoichiometrically deficient in oxygen. Dimanganese heptoxide is an olive-green liquid which separates when powdered $KMnO_4$ is added to concentrated H_2SO_4 at 250 K. Very much the least stable oxide of manganese, it explodes at 285 K, but can be distilled below 270 K.

Black TcO_2 is the most thermally stable oxide of technetium, being formed when either Tc_2O_7 or NH_4TcO_4 is heated. A dihydrate is precipitated when TcO_4^- is reduced with Zn/HCl and the solution is then made alkaline. Rhenium dioxide results from heating Re_2O_7 with Re at 900 K or, as the dihydrate, from a solution of $ReCl_6^{2-}$ treated with OH^-. ReO_2 is less thermally stable than TcO_2, being converted to Re_2O_7 and Re at high temperatures. Both dioxides have the same distorted rutile structure as MnO_2.

TcO_3 has been reported as a product of heating TcO_3Br little is known about it. However ReO_3, a red solid, is well established. It is obtained in almost quantitative yield by the thermal decomposition of $Re_2O_7(C_4H_8O_2)_3$, the addition compound of Re_2O_7 with dioxan. It disproportionates on heating :

$$3\,ReO_3 \rightarrow Re_2O_7 + ReO_2$$

The structure is a very simple one. Every Re atom is octahedrally co-ordinated and every O atom linearly co-ordinated in a cubic lattice. Surprisingly in view of its simplicity, this is not a common lattice, though distorted versions like that in MoO_3 do occur more frequently.

The yellow Tc_2O_7 is driven off from acidic solutions of TcO_4^- on heating Re_2O_7 is not lost at the boiling point of a dilute aqueous solution but does distil from hot solutions of ReO_4^- in strong H_2SO_4. The yellow crystals melt at 490 K ; their structure is interesting – there is an infinite array of ReO_6 octahedra alternating with ReO_4 tetrahedra sharing corners. The lower-melting Tc_2O_7 contains molecules in which a pair of TcO_4 tetrahedra share a corner :

$$O_3Tc \overset{180^\circ}{—O—} TcO_3$$

MANGANATES AND RHENATES

A manganate(V), $Na_3MnO_4(\frac{1}{4}NaOH) \cdot 12\,H_2O$ separates from solution when $KMnO_4$ is reduced with alkaline Na_2SO_3. It is thermally unstable, but more stable salts such as $Ba_3(MnO_4)_2$ and K_2MnO_4 have also been prepared.

The well-known green MnO_4^{2-} ion is formed when MnO_2 is fused with an alkali-metal hydroxide and an oxidising agent such as KNO_3. The salts are unstable to disproportionation in even slightly acidic solution :

$$3\,MnO_4^{2-} + 4\,H^+ \rightarrow 2\,MnO_4^- + MnO_3 + 2\,H_2O$$

One of the most stable manganates(VI), $BaMnO_4$, can be made by adding concentrated aqueous $KMnO_4$ to boiling, saturated baryta water:

$$4\,Ba(OH)_2 + 4\,MnO_4^- \rightarrow 4\,OH^- + 2\,H_2O + O_2 + 4\,BaMnO_4$$

The room-temperature magnetic movement of $BaMnO_4$ is $1.80\mu_B$.

Rhenates(VI) of various types have been made by fusing together a per-rhenate and rhenium with alkali-metal oxides. The best known are those containing the ReO_6^{6-} ion; Ca_3ReO_6, and the corresponding Sr and Ba salts have all been made.

PERMANGANATES, PERTECHNETATES AND PERRHENATES

Permanganates

The stability of the permanganate ion MnO_4 over a wide range of pH makes the permanganates very useful oxidising agents. Potassium permanganate. $KMnO_4$, is manufactured by the electrolytic oxidation of the alkaline manganate solution. It is used in volumetric oxidimetry, in the industrial production of such things as saccharin and benzoic acid, and for bleaching waxes. The MnO_4^- anion is much used in the volumetric determination of manganese since Mn^{2+} is quantitatively oxidised to it is dilute nitric acid by insoluble oxidising agents like sodium bismuthate.

In alkali MnO_4^- reacts as an oxidising agent thus :

$$MnO_4^- + 2\,H_2O + 3\,e \rightarrow MnO_2 + 4\,OH^- \qquad E^\circ = +1.23\ V$$

and in acid oxidation occurs thus :

$$MnO_4^- + 8\,H^+ + 5\,e \rightarrow Mn^2 + 4\,H_2O \qquad E^\circ = +1.51\ V$$

A solution of $HMnO_4$ can be made by adding H_2SO_4 to aqueous $Ba(MnO_4)_2$ below 272 K. A hydrate $HMnO_4 \cdot 2\,H_2O$ has been isolated from the solution. The acid is very soluble in water and ionises strongly.

Permanganates exhibit weak, temperature-independent magnetism which is due to a second-order Zeeman effect between the higher molecular orbital levels and the ground level.

Pertechnetates and perrhenates

A colourless solution of $HTcO_4$ is obtained when yellow Tc_2O_7 is dissolved in water, but as the solution is concentrated it becomes pink and then red ; finally dark-red crystals of $HTcO_4$ separate from solution. This colour phenomenon is not yet explained. Colourless $KTcO_4$ has high thermal stability but NH_4TcO_4 decomposes to give TcO_2 :

$$2\,NH_4TcO_4 \rightarrow N_2 + 2\,TcO_2 + 4\,H_2O$$

Solution of $HReO_4$, which are also colourless, are made by dissolving Re_2O_7 in water or oxidising either Re itself or ReO_2 with aqueous H_2O_2. On evaporation the solution becomes greenish-yellow; careful drying over P_2O_5 eventually yields a yellow, hygroscopic solid which is a true molecular hydrate, $Re_2O_7 \cdot 2\,H_2O$. Line $HMnO_4$ and $HTcO_4$, aqueous $HReO_4$ is a strong acid. The only anion present in the solution is the tetrahedral ReO_4^- ion, but heavy-metal ions precipitate mesoperrhenates, $M_3^{II}(ReO_5)_2$. The acid is not a particularly strong oxidising agent but is oxidises HBr to Br_2.

The alkali-metal perrhenates are colourless, thermally stable, and except for $NaReO_4$, rather insoluble $KReO_4$, made by adding KCl to aqueous $HReO_4$, is less soluble than $KClO_4 \cdot NH_4ReO_4$ decomposes on strong heating, the products being ReO_2, N_2 and H_2O.

SULPHIDES

Manganese(II) sulphide has three forms. The stable, green α–MnS, alabandite, has the NaCl structure, but there are metastable red forms with the blende and wurtzite structures respectively. The compound MnS_2 contains Mn^{2+} and S_2^{2-} ions in a structure closely related to iron pyrites. MnS_2 and all three forms of MnS are semiconductors and are antiferromagnetic at low temperatures. Technetium and rhenium form the isomorphous pairs TcS_2, ReS_2 and Tc_2S_7, Re_2S_7. The black heptasulphides are precipitated when H_2S is passed under pressure into acid solutions of TcO_4^- and ReO_4^-. The sulphur which is co-precipitated is removed by leaching with CS_2. The disulphides are obtained from the heptasulphides by heating in vacuo.

BINARY COMPOUNDS WITH OTHER NON-METALS

Manganese combines with nitrogen at 1000 K to give a mixture of nitrides ; the composition of the product depending on the pressure of the gas. Rhenium does not react with N_2 to give a nitride, but NH_4ReO_4 reacts with H_2 at 580 K to give a mixture of the metal with Re_3N.

The Mn–C system is complicated, and the reactions are difficult to study because manganese is so reactive towards oxygen and nitrogen. Orthorhombic Mn_3C hydrolyses to give a mixture of gaseous products, typically 75% H_2, 15% CH_4 and 10% C_2H_6. There is some doubt about the existence of rhenium carbides.

Both manganese and rhenium form mixtures of borides by direct combination at high temperature. They resemble other transition-metal borides in appearance and properties.

POTASSIUM ENNEAHYDRIDORHENATE(VII)

When potassium perrhenate is reduced with a solution of potassium in slightly moist ethylenediamine, colourless, diamagnetic, hexagonal crystals are obtained. The positions of the Re and K atoms were found by X-ray diffraction and those of the H atoms by neutron

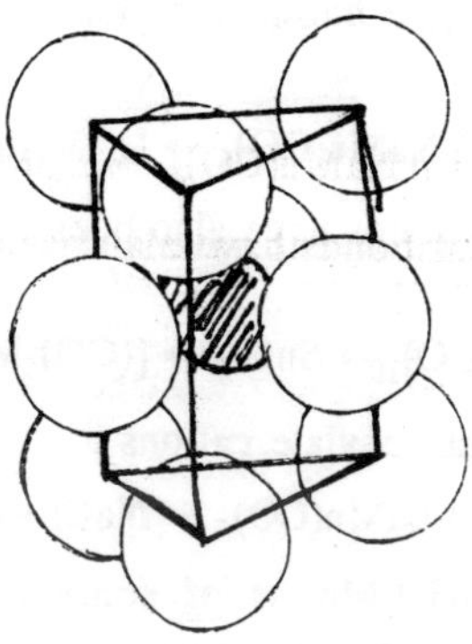

Fig. 5.7. The ReH_9^{2-} ion.

diffraction. Proton magnetic resonance shows a single peak at high field indicating equivalence or vibrational deformation of the H atoms. The anion, ReH_9^{2-}, has the structure shown in Fig. 5.7. The Re atom at the centre of a triangular prism is surrounded by 9 H atoms, six at the corners and three on extensions of perpendiculars through the centres of the vertical sides. The Re–H distances are all 168 pm. If the hydrogens are considered to be co-ordinated as H^- ions, the Re is in a formal oxidation state of +7, and the diamagnetism is thereby explained. In the crystal the K^+ ions lie in straight lines through the H and Re atoms and form a larger tricapped triangular prism at 60 about a common vertical axis to the hydrogen one.

ORGANOMETALLIC COMPOUNDS AND π-COMPLEXES

Only one binary metal carbonyl of manganese is known, $Mn_2(CO)_{10}$, best made by the action of CO under pressure of MnI_2 in the presence of a reducing agent such as an alkyialuminium. The golden-yellow crystals are stable in air at room temperature but decompose quickly above 380 K. The molecule belongs to the symmetry group D_{4d} as confirmed by its vibrational spectrum. The carbonyl groups have been substituted by a variety of ligands usually to give compounds of the types $Mn_2(CO)_8L_2$ and $Mn_2(CO)_9L$.

Sodium pentacarbonylmanganate(–I), $NaMn(CO)_5$, made by the reduction of $Mn_2(CO)_{10}$ with sodium amalgam in tetrahydrofuran, is a useful starting material for the preparation of many compounds containing bonds between Mn and other metals such as Ge, Pb, Mo, Fe, Co, Cu, Ag and Hg which are a particular feature of the chemistry of manganese. Typically, a halide derivative of other metal reacts with $NaMn(CO)_5$:

$$L_m MX_n + n\, NaMn(CO)_5 \rightarrow L_m M[Mn(CO)_5]_n + n\, NaX$$

Manganese–metal bonds have also been produced by insertion reactions, *e.g.* :

$$Mn_2(CO)_{10} + SnCl_2 \rightarrow [(CO)_5Mn]_2SnCl_2$$

and by reactions with carbonylate cations :

$$NaCo(CO)_4 + CiMn(CO)_5 \rightarrow NaCl + (CO)_4CoMn(CO)_5$$

Manganocene, $(C_5H_5)_2Mn$, is of some interest. The number crystals can be made by treating manganese dihalides in tetrahydrofuran with C_5H_5Na. X-ray studies show the molecule to have a sandwich structure like ferrocene and magnetic studies show the magnetic moment to be $5.86 \pm 0.05\mu_B$, consistent with the presence of an Mn^{2+} (d^5) ion. Thus the indications are that the metal–ring bonds are ionic ; the Mn compound is unique among d-block metal-cyclopentadienyl compounds in this respect ; in all the others the metal–ring bonds are covalent. Manganocene reacts with CO under pressure to give the yellow solid $C_5H_5Mn(CO)_3$ in which the C_5H_5 undergoes attack by some electrophiles to give organic derivatives in much the same way as ferrocene does. Substitution of the CO groups can be effected under ultraviolet irradiation by amines, phosphines and arsines as well as olefins and acetylenes. In $(C_5H_5)Mn(CO)_2PhC \equiv CPH$ the metal is bonded by π-electrons of the acetylenic bond occupying the position vacated by the CO group.

Tc and Re form carbonyls $M_2(CO)_{10}$ with structures like that of manganese carbonyl ; $Re_2(CO)_{10}$ is obtained in good yield by the action of CO on Re_2O_7 at 520 K and 20 MPa without the need for an additional reducing agent.

The reaction between C_5H_5Na and $ReCl_5$ in tetrahydrofuran yields not $(C_5H_5)_2$ Re but the hydride $(C_5H_5)_2ReH$. $TcCl_4$ reacts similarly with C_5H_5Na to give $(C_5H_5)_2$ TcH. The rhenium compound has been carefully studied. It is diamagnetic and basic, its hydrogen being hydridic, not protonic. The lemon-yellow crystals dissolve in dilute HCl to give a salt :

$$(C_5H_5)_2ReH + HCl \rightarrow [(C_5H_5)_2ReH_2]^+Cl^-$$

It was in this compound that the characteristics of a bond between a transition-metal atoms and a hydrogen atom were first recognised.

At 370 K and 25 Mpa, $(C_5H_5)_2ReH$ reacts with CO to give pale-yellow crystals of composition $C_{12}H_{11}O_2Re$ which have been shown by i.r. and n.m.r. studies to contain molecules with the structure:

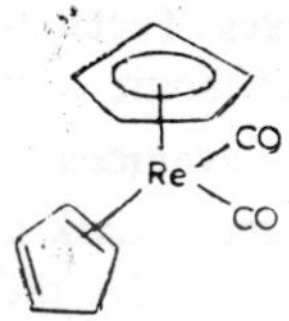

Interestingly, the compound obtained by the action of $Re_2(CO)_{10}$ on C_5H_6 is quite different ; it is π–$C_5H_5Re(CO)_3$.

COMPLEXES

The +1 state is represented by hexacyano-complexes, $K_5M(CN)_6$ for all three metals. The preparation of the olive-green $K_5Tc(CN)_6$ is typical ; it is made by reducing TcO_4^- with potassium amalgam in the presence of C N. Exposure to air converts it to the Tc^{IV} complex $[Tc(OH)_3(CN)_4]^{3-}$.

The +2 state is common in manganese, rather uncommon in rhenium and rare in technetium. The d^5 configuration is rather unfavourable for the formation of low-spin octahedral complexes but they exist in $Mn(CN)_6^{4-}$ and $MN(CNR)_6^{2+}$. The high-spin ammines of Mn^{II} are rather unstable, but there are octahedral complexes with chelating ligands such as ethylenediamine and the oxalate ion. Mn^{II} has some tetrahedral complexes, mainly salts of MnX_4^{2-} (X = Cl, Br,I) with large cations such as Me_4N^+.

Tc^{II} and Re^{II} exist in the compounds $MCl_2(diarsine)_2$, made by reducing M^{VII} compounds with hypohosphite in methanol in the presence of Cl^- ions and 0–$C_6H_4(AsMe_2)_2$.

The 3+ state is moderately common in this sub-group. Several ligands lower $E°$, Mn^{III}/Mn^{II} sufficiently for Mn^{II} to be oxidised in air. Examples of these are CN^- in $K_3Mn(CN)_6$, and acetylacetone in $Mn(acac)_3$, both of which can be made by atmospheric oxidation of solutions of Mn^{2+} and the ligand. Tc^{III} and Re^{III} compounds $[MX_2(diarsine)_2]\,ClO_4$ have been made by a method similar to that for the foregoing diarsine complexes, but with the use of weaker reducing agents.

The +4 state for manganese is not common – it exists in some hexachloroand hexafluorocomplexes. Technetium(VI), in addition to its halide complexes, forms some cyanocomplexes. Thus $TcO_2 \cdot 2\,H_2O$ dissolves in alkali cyanides to give $Tc(OH)_3(CN)_4^{3-}$ which can be isolated as the dark-brown thallium salt. K_2TcI_6 and KCN react in methanol to give the dark-read $K_2Tc(CN)_6$. But similar reactions with the corresponding rhenium compound produce the Re^V complexes $re(OH)_4(CN)_4^{3-}$ and $K_3Re(CN)_8$.

The +5 state in technetium is represented by $[TcCl_4(diars)_2ClO_4$; this provided the first example of 8-co-ordinate technetium.

Higher oxidation states than +5 in this sub-group are represented mainly by the oxo-complexes and complex halides. But another particularly interesting complex is the rhenium(VI) compound tris(cis-1, 2-diphenylethene-1, 2-dithiolato)-rhenium :

$$Re\left(\begin{array}{l} S-C-Ph \\ \quad\;\, \| \\ S-C-Ph \end{array}\right)_3$$

CHAPTER 6

Iron, Cobalt and Nickel

INTRODUCTION

The nine elements of Group VIII carry the three d-block series from the manganese group, VIIA, to the copper group, IB. Although there are the usual vertical similarities, as for example in Fe, Rh and Os, the horizontal similarities, as in Fe, Co and Ni, are particularly strong. The metals Fe, Co and Ni replace hydrogen from non-oxidising acids, form aquated ions with charges of either +2 or +3, and behave, in general, as typical active metals. But the heavier congeners, Ru, Rh, Pd, Os, Ir and Pt, are generally unreactive towards acids, though Pd is converted to $Pd(NO_3)_2$ and Os to OsO_4 by concentrated HNO_3, and their typical ions are anionic complexes such as $PtCl_6^{2-}$ in which the metal atom is in a high oxidation state and is covalently bound to the ligands which surround it. These metals, often known collectively as the platinum metals, are very similar in most physical and chemical properties. To simplify the correlation of properties in Group VIII, the active metals, Fe, Co and Ni will be treated first and the platinum metals will be considered in the next chapter.

THE ELEMENTS

The three metals resemble Cr and Mn, which precede them, in their physical properties and reactivity. All three combine with oxygen on heating; finely divided Fe and Ni are, in fact, pyrophoric. The metals also combine with the halogens, sulphur, boron, carbon, silicon and phosphorus. It is known that carbide, silicide and nitride phases are important in the metallurgy of iron. Nickel is the only transition element which combines with carbon monoxide at atmospheric pressure, tetracarbonyl nickel, $Ni(CO)_4$, being formed at 325 K. Iron and nickel dissolve easily in dilute, non-oxidising acids, but with cobalt reaction is slow even though E° , Co^{2+}/Co is negative (Table 6.2). Hot HNO_3 renders all three metals passive. Iron is oxidised quickly in moist air, the hydrated oxide which is formed being of no protection to the underlying metal because it flakes off easily. Co and Ni resist atmospheric corrosion at ordinary temperatures, however, and nickel plating is used as a protective

TABLE 6.1 ATOMIC PROPERTIES OF Fe, Co AND Ni

	Fe	Co	Ni
Z	26	27	28
Electron configuration	$[Ar]3d^64s^2$	$[Ar]3d^74s^2$	$[Ar]3d^84s^2$
$I(1)$/kJ mol^{-1}	762	758	736
$I(2)$/kJ mol^{-1}	1561	1644	1752
Metallic radius/pm	126	125	124
rM^{2+}/pm	76	74	72
rM^{3+}/pm	64	63	

TABLE 6.2 PHYSICAL PROPERTIES AND REDOX POTENTIALS

	Fe	Co	Ni
M.p./K	1808	1760	1728
ρ/g cm^{-3}	7.9	8.9	8.9
E°,M^{2+}/M/V	– 0.44	– 0.27	– 0.24
E°,M^{3+}/M^{2+}/V	+ 0.77	+ 1.84	

coating on steel. The metals all low-temperature α-forms, b.c.c. in the case of Fe and h.c.p. in Co and Ni, and also high-temperature modifications, γ-Fe, β-Co and β-Ni, all with the c.c.p. lattice. All three metals are ferromagnetic; the Curie temperature for Co is particularly high, near 1400 K.

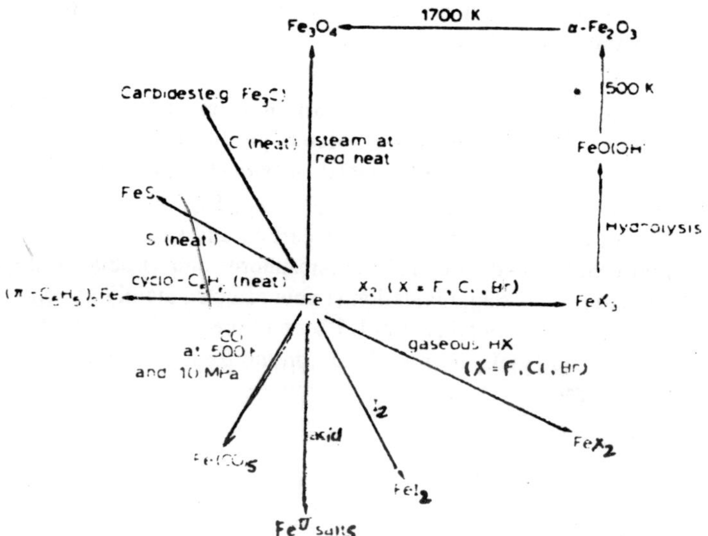

Fig. 6.1. An outline of the chemistry of iron.

THE ELEMENTS

Iron

Iron (5.1% of the lithosphere) is widespread, many minerals owning their tints to its presence, but the important ores are haematite, Fe_2O_3, magnetite, Fe_3O_4, and siderite $FeCO_3$. Iron pyrites, FeS_2, is a common vein mineral, but is not worked up for the metal because its sulphur content is so high. The core of the Earth is thought to consist mainly of iron, and the metal is the major constituent of metallic meteorites.

Pure iron, which can be made by hydrogen reduction of the oxides, has little industrial use, but iron alloys are of great importance. Annual world production probably exceeds 300 million tons, which represents more than ten times the total of all other metals. The first stage in the industrial production is to reduce the calcined ores (Fe_2O_3) with coke in a blast furnace. Limestone is used to remove silica as slag. In the cooler part of the furnace the reducing agent is CO, and typical reactions are :

$$3\,Fe_2O_3 + CO \rightarrow 2\,Fe_3O_4 + CO_2$$

$$Fe_3O_4 + 4CO \rightarrow 3\,Fe + 4\,CO_2$$

$$Fe_2O_3 + CO \rightarrow 2\,FeO + CO_2$$

In the lower, hotter parts of the furnace (> 1800K) the reduction is mainly by carbon:

$$FeO + C \rightarrow Fe + CO$$

Most of the silica in the original ore is tapped off as a lag calcium silicate, but some silicon, produced by the reduction of SiO_2, is incorporated in the metal. A typical pig iron, tapped from the furnace, contains about 4% C, 2.5% Si, 2.5% Mn, 1% P and 0.1% S. To convert pig iron to steel, the carbon and silicon must be largely removed by oxidation, and the phosphorus and sulphur by the use of lime to form a slag. Oxidation is now effected mainly by oxygen gas. Steel usually contains 0.2–1.5% combined carbon, which must be added to the melt, after oxidation, in the form of carbides of iron and manganese.

Cobalt

Cobalt (10^{-3}% of the lithosphere) is widely distributed, but its workable ores are mainly sulphides and arsenides which are associated with other metals, particularly nickel and copper. Examples are smaltite, $CoAs_2$, and cobaltite, CoAsS. The extraction of cobalt is usually ancilliary to the production of the other metal in the ore, but in a typical process the roasted ore is dissolved in

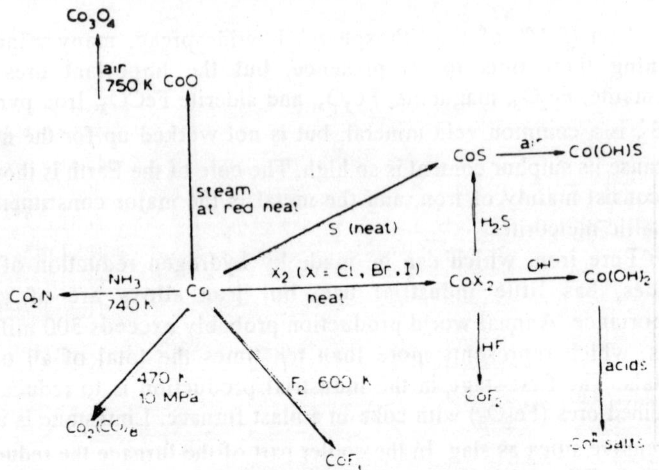

Fig. 6.2. An outline of cobalt chemistry.

H2SO4, iron is removed with lime and cobalt is precipitated as CoO(OH) by addition of NaOCl. Heating converts the precipitate to Co3O4, which can be reduced with carbon.

Nickel

Nickel (0.016% of the lithosphere) occurs mainly as sulphides and arsenides. The important deposit at Sudbury, Ontario is mainly pentlandite, (Ni,Fe)S. The concentrate is roasted and smelted to form a matte, which, on heating with $NaHSO_4$ and coke, separates into two layers, the lower being mainly NiS. This is oxidised to NiO which is used for making nickel steels. The pure metal is obtained by the Mond process; first crude nickel is obtained by reducing NiO with water gas, this metal is then converted to volatile $Ni(CO)_4$ by the action CO at 330 K, finally the $Ni(CO)_4$ is decomposed by passing it through nickel pellets at 450 K, the CO which is released being recycled over the impure metal.

OXIDATION STATES

The lowest oxidation states occur as usual in π-bonded complexes such as carbonyls and carbonylate anions, and the highest in fluorocomplexes, or in the case of Fe, in an oxoanion (Table 6.3). However, the highest oxidation state falls progressively with atomic number from the value of +7 reached in manganese.

TABLE 6.3 REPRESENTATIVE COMPOUNDS AND IONS OF OXIDATION STATES –2 TO +6 OF IRON, COBALT AND NICKEL

Charge	Fe	Co	Ni
–2	$Fe(CO)_4^{2-}$	–	–
–1	–	$Co(CO)_4^-$	$Ni_2(CO)_6^{2-}$
0	$Fe(CO)_5$	$K_4Co(CN)_4$	$Ni(PF_3)_4$, $K_4Ni(CN)_4$
+1	$(\pi\text{-}C_5H_5)_2Fe_2(CO)_4$	$Co(dipy)_3^+$, $Co(NCC_6H_5)_5ClO_4$	$K_4Ni_2(CN)_6$
+2	$Fe(H_2O)_6^{2+}$, $FeCl_2$,, $(\pi\text{-}C_5H_5)^2Fe$	$Co(H_2O)_6^{2+}$, $CoCl_4^{2-}$	$Ni(H_2O)_6^{2+}$, $Ni(en)_3^{2+}$
+3	FeF_6^{3-}, $FeO(OH)$, $K_3Fe(CN)_6$	$Co(NH_3)_6^{3+}$, $K_3Co(CN)_6$	$Ni(diars)_2Cl_2^+$, $NiBr_3(PMe_3)_3$
+4	$Fe(diars)_2Cl_2^{2+}$, Ba_2FeO_4	CoF_6^{2-}	K_2NiF_6
+5	FeO_4^{3-}	–	–
+6	K_2FeO_4	–	–

Zero oxidation state is particularly common in Ni, for example the carbonyl groups in $Ni(CO)_4$ can be replaced by other neutral π-acceptor ligands such as isonitriles and phosphorus trihalides. The charge number +1 is uncommon, but all three elements in their dipositive states form very large numbers of compounds; Ni^{II} often forms square complexes which are diamagnetic, but high-spin tetrahedral and also octahedral complexes are common.

The terpositive state occurs much more in Fe and Co than in Ni. The Fe^{3+}aq ion is a moderately strong oxidising agent $E°, Fe^{3+}/Fe^{2+} = +0.77$ V) and the Co^{3+}aq ion a powerful one $E°, Co^{3+}/Co^{2+} = +1.84$ V) but in coordination with ligands of only moderate field strength, like ammonia and ethylenediamine, Co^{III} (d^6) forms diamagnetic complexes, many of considerable thermodynamic stability.

Charge +4 is rather uncommon, but occurs in some diarsine complexes of iron and in CoF_6^{2-} and NiF_6^{2-}. Charge +6 occurs in the ferrates (VI) which are made by the oxidation of a suspension of hydrated iron (III) oxide in aqueous alkali:

$$Fe_2O_3 + 3\,OCl^- + 4OH^- \rightarrow 2\,FeO_4^{2-} + 3Cl^- + 2\,H_2O$$

The carmine-red $BaFeO_4$ is similar to $BaCrO_4$ in structure; though fairly stable to reduction in aqueous alkali, the FeO_4^{2-} ion is a stronger oxidising agent than MnO_4^- in acidic solutions.

HALIDES

White FeF_2 can be made by passing HF over iron at red heat or over $FeCl_2$ at a lower temperature. It has the rutile structure. The yellow dichloride and greenish-yellow dibromide, made by passing HCl or HBr over the strongly heated metal, both have layer lattices, $FeCl_2$ the $CdCl_2$ structure, and $FeBr_2$ the CdI_2 structure. The iodide, which also has the CdI_2 lattice, can be made from iron and iodine. The halides form many hydrates. Interestingly, $FeCl_2 . 6\,H_2O$ has been shown to contain *trans*-$FeCl_2(H_2O)_4$ units and not hexa-aquoiron (II) ions.

There are three trihalides. Green FeF_3, which is a rather insoluble compound, can be made by the action of fluorine on the metal. The very soluble, black $FeCl_3$ can also be made by direct combination, but $FeBr_3$ is difficult to obtain pure because it decomposes rather easily to $FeBr_2$ and bromine. The chloride and bromide have layer structure similar to that of BiI_3 (26.3.3).

Cobalt forms all four dihalides. CoF_2, best made by heating $CoCl_2$ in a stream of HF, is a pink solid with the rutile structure. The other three dihalides, which are unlike CoF_2 in being readily soluble

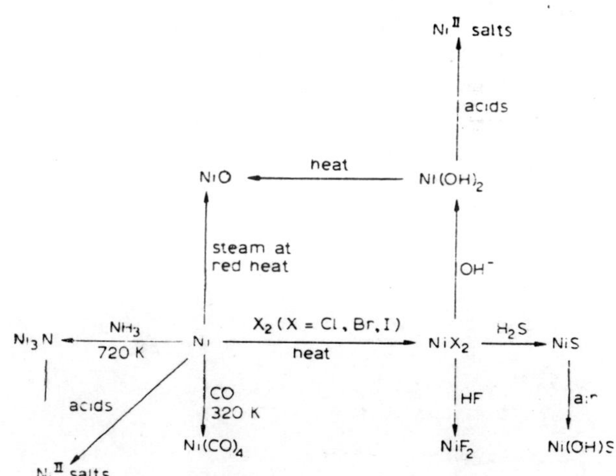

Fig. 6.3. An outline of nickel chemistry.

Fig. 6.3. An outline of nickel chemistry.

in water, and in having layer lattices, can be made by direct combination of the elements. There are several hydrated halides. Blue $CoCl_2$ turns pink in moist air and it is used as an indicator in desiccants such as silica gel. The pink hexahydrate contains *trans*-$CoCl_2(H_2O)_4$ units. A violet dihydrate, $CoCl_2 \cdot 2\,H_2O$, crystallises from aqueous solution above 325 K, and a blue-violet monohydrate at 363 K.

Anhydrous CoF_3 can be made as a light-brown powder by treating Co with F_2 at 500 K, but it has not been possible to prepare the other trihalides.

Green, hydrated nickel dihalides can be crystallised from solutions of the metal, its oxide or carbonate in the appropriate aqueous hydrogen halides. The anhydrous chloride and bromide, both yellow solids with the $CdCl_2$ structure can be made by direct combination of the elements. The yellow NiF_2, which has the rutile structure, is best made by treating $NiCl_2$ with HF at 750 K, and the black NiI_2 results from the mixing of ethanolic solutions of NaI and $NiCl_2$. Its hexahydrate contains $Ni(H_2O)_6^{2+}$ ions, unlike $NiCl_2 \cdot 6\,H_2O$ and $NiBr_2 \cdot 6\,H_2O$ which have *trans*-$NiX_2(H_2O)_4$ units.

COMPLEX HALIDES

Complex anion of general formula MX_4^{2-} are very common. Fe^{II} and Ni^{II} halides react in ethanol with salts such as quaternary ammonium halides to give compounds $(R_4N)_2MX_4$, for example. The CoX_4^{2-} complexes are even easier to prepare: the chloro-, bromo- and iodo-complexes will crystallise from aqueous mixtures of CoX_2 with MX (M = alkali metal). The MX_4^{2-} ions are all tetrahedral or very nearly so.

High oxidation states are common in the fluorocomplexes. Yellow Cs_2CoF_6, for example, is obtained by treating Cs_2CoCl_4 with fluorine at 550 K. Red K_2NiF_6 is made in similar way; a mixture of $2\,KCl + NiCl_2$ is fluorinated. The diamagnetic compound liberates oxygen from water.

CYANIDES AND CYANOCOMPLEXES

The hexacyanoferrate(II) ion, which is non-poisonous, occurs with many cations; the potassium salt $K_4Fe(CN)_6 \cdot 3\,H_2O$ can be crystallised from the solution after treating an Fe^{II} salt with an excess of KCN. Many oxidising agents convert it to hexacyanoferrate (III), e.g. :

$$Ce^{4+} + Fe(CN)_6^{4-} \rightarrow Ce^{3+} + Fe(CN)_6^{3-}$$

Hexacyanoferrate(II) reacts with an excess of Fe^{3+} to give the pigment Prussian blue, whereas $Fe(CN)_6^{3-}$ reacts with Fe^{2+} to give a precipitate known as Turnbull's blue. The two compounds have been found to give the same X-ray powder diffraction pattern and the same Mossbauer spectrum. If 1 : 1 molar proportions of Fe^{3+} and $Fe(CN)_6^{4-}$ react together, soluble Prussian blue is obtained. This has the formula $KFe[Fe(CN)_6] \cdot H_2O$. The electronic and Mossbauer spectra indicate a formulation $KFe^{III}[Fe^{II}(CN)_6]$ in which high-spin Fe^{3+} ions are octahedrally surrounded by N atoms and low-spin Fe^{2+} by C atoms (Fig. 6.4.).

Unlike iron (II) cyanide, $Co(CN)_2$ can be made from the dipositive ion and CN^- in aqueous solution. When the compound is dissolved in aqueous KCN and ethanol is added, a violet salt $K_6[Co_2(CN)_{10}] \cdot 4\,H_2O$ is precipitated. Its anion exhibits three strong bands in the CN stretching region, and is thought to be structurally similar to $Mn_2(CO)_{10}$.

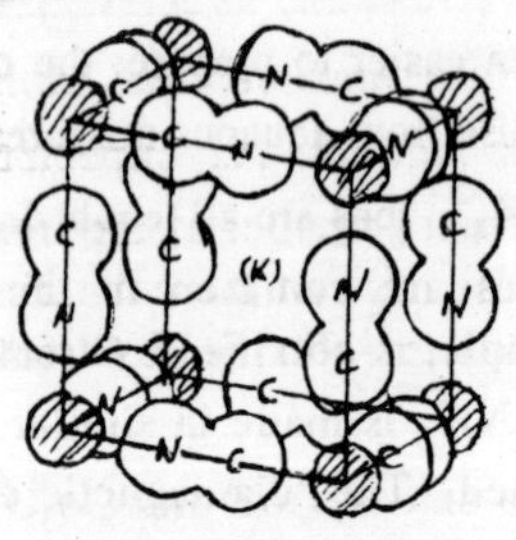

Fig. 6.4. Unit cell of soluble Prussian blue. K^+ ion in alternate cells.

$Ni(CN)_2$ can be precipitated as a grey hydrate from Ni^{2+} solutions treated with CN^-. It dissolves in an excess of aqueous KCN, and potassium tetra-cyanonickelate(II), in the form of an orange monohydrate, can be crystallised from the solution. The diamagnetic compound contains a square $Ni(CN)_4^{2-}$ ion. It can be reduced with potassium in liquid ammonia to potassium tetra-cyanonickelate(0), $K_4Ni(CN)_4$; a yellow solid which liberates hydrogen from water.

OXIDES

There are three oxides of iron, with the ideal composition FeO, Fe_3O_4, Fe_2O_3, but they all show gross nonstoichiometry. The black powder made by heating iron(II) oxalate in the absence of air at ordinary pressure is always iron-deficient, its iron-rich limit being $Fe_{0.95}O$. However, it is claimed that stoichiometric FeO can be made at 5 GPa.

The brown compound FeO(OH), which is made by hydrolysis of $FeCl_3$ at high temperature, is converted at 470 K into red-brown α–Fe_2O_3 ; the same compound occurs in nature as haematite. This has the corundum structure in which the oxide ions are arranged as in hexagonal close-packing but there is another form, γ–Fe_2O_3, in which the oxide ions are arranged as in cubic close-packing. The black compound Fe_3O_4, which occurs as magnetite, can be made in the laboratory by heating Fe_2O_3 above 1700 K.

FeO, Fe_3O_4 and γ–Fe_2O_3 are closely related structurally. In FeO most of the octahedral spaces between the oxide ions are filled with Fe_{2+} ions in a defect NaCl lattice. In Fe_3O_4 there are both Fe^{2+} and Fe^{3+} ions in an inverse spinel structure and in γ–Fe_2O_3 there are Fe^{3+} ions but not Fe^{2+} ; however the cubic arrangement of O^{2-} ions is preserved throughout. As expected, the cubic unit containing 32 O^{2-} ions becomes smaller as the iron ions are reduced in both number and size. The edge of the cube diminishes from 860 pm to 828 pm as the atomic ratio Fe : O falls from 0.486 in $Fe_{0.95}O$ to 0.400 in Fe_2O_3.

Cobalt forms only two oxides, CoO and Co_3O_4. The olive-green CoO, conveniently made by heating $CoCO_3$, has the NaCl structure and a composition very close to the ideal one. It is antiferromagnetic at room temperature. when heated to 700 K in oxygen it is converted to the normal spinel Co_3O_4. The green nickel(II) oxide, made by heating $NiCo_3$ or $Ni(NO_3)_2$, also has the NaCl lattice and almost ideal composition. It has been suggested that the very narrow variation of composition in CoO and NiO compared with FeO and VO is due to the greater effective nuclear charge acting on the outer electrons of the dipositive ions, making them less easy to promote to nonlocalised orbitals. There is no evidence for a nickel (III) oxide, but a well-defined compound, β–NiO(O H), is obtained as a black precipitate by oxidising aqueous Ni^{2+} with potassium hypobromite solution. A similar compound of cobalt is similarly made.

The oxides of iron, cobalt and nickel form compounds with oxides of other metals; the ferrites, which are dealt with in the following paragraph, are of particular technological interest.

FERRITES

These compounds, which are made by heating the appropriate metal carbonate with Fe_2O_3, have the general formula $M^{II}Fe_2^{III}O_4$. The normal ferrites, such as $ZnFe_2O_4$ and $CdFe_2O_4$, with normal spinel structure, are diamagnetic; the inverse ferrites, with inverse spinel structure, exhibit *ferrimagnetism.* In ferrimagnetic spinels all the ions occupying tetrahedral sites have parallel electronic spins and all the unpaired ions occupying octahedral sites have parallel spins, but the second set are antiparallel to the first set. Thus at low temperatures the inverse ferrite $Fe^{III}(Fe^{III}Ni^{II})O_4$ exhibits the magnetic susceptibility arising from two unpaired spins (i.e. 7–5) per formula unit :

The ferrimagnetism of an inverse ferrite can be enhanced by lattice substitution. In a mixed ferrite of composition $Zn^{II}_{\frac{1}{2}}Ni^{II}_{\frac{1}{2}}Fe_2O_4$, the distribution of ions among tetrahedral sites will be :

$$(Zn^{II}{}_{1/2}\,Fe^{III}{}_{1/2})_{tetr}\,(Fe^{III}{}_{3/2}Ni^{II}{}_{1/2})_{oct}\,O_4$$

Thus the number of unpaired spins per formula unit might be expected to increase to six (i.e. $7\frac{1}{2}+1-2\frac{1}{2}$). In practice this enhancement of the magnetic susceptibility proceeds to only a limited extent because the exchange forces are weakened as the diamagnetic Zn^{2+} ions are added (Fig. 6.5).

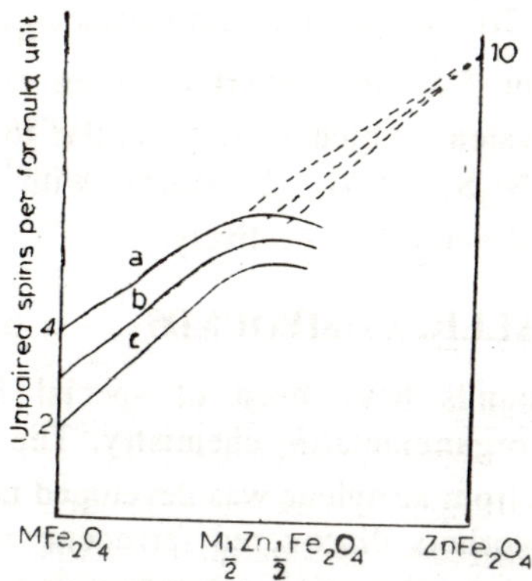

Fig. 6.5. Variation of number of unpaired spins per formula unit for mixed ferrites. In (a) M = Fe, in (b) M = Co and in (c) M = Ni.

The employment of ferrites in electromagnetic devices is associated with (a) their very square hysteresis loops, (b) their high-speed response, and (c) their high flux output. Thus the direction of an induced magnetic field in a ferrite can be sharply reversed by an electric impulse of the right size, but remains unaffected by a smaller impulse. This property is used particularly in the storage and retrieval of information, using binary notation, in the memory banks of computers.

SULPHIDES

The iron-sulphur system has been extensively studied but in not yet fully characterised. FeS_2 occurs in two forms, pyrites and marcasite, both brassy in appearance. Pyrites has a cubic lattice, with Fe^{2+} and S_2^{2-} ions arranged like the ions in rock salt. The S–S bonds are parallel to trigonal axes. Both forms of FeS_2 are diamagnetic, indicating that the Fe^{2+} ions are spin-paired ($t^6{}_{2}g$). Grey FeS, which has the nickel arsenide structure, is made by direct combination. It also occurs as pyrrhotite, which is usually iron-deficient; a phase Fe_7S_8, which has an NiAs structure with one-eighth of the metal positions unoccupied, is a ferromagnetic material.

In the cobalt–sulphur system the compounds which are identified are CoS_2, with the pyrites structure, Co_3S_4, with a spinel structure, and the metal-deficient $Co_{1-x}S$. CoS_2 is ferromagnetic below 120 K ; its paramagnetic moment at room temperature indicates that the Co^{2+} ion has the spin-paired $t^6_{2g}e^1_g$ configuration an indication that the S_2^{2-} ions exert a strong crystal field. In the nickel–sulphur system, which is very similar to the Co–S system, there are NiS_2, Ni_3S_4 and $Ni_{1-x}S$ phases with pyrites, spinel and nickel aresenide structures respectively.

ORGANOMETALLIC COMPOUNDS

Iron compounds have been of special importance in the development of organometallic chemistry. The use of $Fe(CO)_5$ in organic synthesis from acetylene was developed in 1949, and in 1951 two groups of workers discovered ferrocene and began the now extensive study of organometallic compounds containing π-bonded aromatic rings. Ferrocene, $(C_5H_5)_2Fe$, was made (a) by treating cyclopentadienyl magnesium bromide with $FeCl_3$ and (b) by passing cyclopentadiene with nitrogen at 570 K over finely divided iron:

$$2\,C_5H_6 + Fe \rightarrow (C_5H_5)_2Fe + H_2$$

It is now made most conveniently by treating $FeCl_2$ with sodium pentadienide in tetrahydrofuran or diglyme:

$$FeCl_2 + 2\,C_5H_5Na \rightarrow (C_5H_5)_2Fe + 2\,NaCl$$

The orange crystals are insoluble in water but soluble in most organic solvents. X-ray diffraction studies show that in the crystal the iron atom is sandwiched between two C_5H_5 rings which are staggered relative to one another. All Fe–C distances are the same. In the vapour the two rings are eclipsed, as shown by electron-diffraction studies.

To construct a molecular orbital scheme to explain the bonding we must first consider the C_5H_5 rings. In these the five 2p π atomic orbitals combine to form five π molecular orbitals which fall into three groups. The MO of lowest energy is one of symmetry designation A_1, there are two degenerate E_1 orbitals with a nodal plane perpendicular to the ring, and those of highest energy are the E_2 orbitals with two such nodal planes. These localised molecular orbital are then combined to give a set of ten molecular orbitals encompassing both rings; these molecular orbitals have symmetry classifications A_{1g}, A_{2u}, E_{1g}, E_{1u}, E_{2g}, and E_{2u}. In D_{5d} symmetry, the atomic orbitals of the iron atom, the 3d, 4s and 4p, have the following representations: $A_{1g}(4s, 3d_{z^2})$, $A_{2u}(4p)$, $E_{1g}3d_{xz}, 3d_{yz})$, $E_{2g}(3d_{xy}, 3d_{x^2-y^2})$

and $E_{1u}4p_x$, $4p_y$). As in all MO constructions, only those orbitals with the same symmetry properties can give rise to net overlap. Calculation shows that twelve bonding electrons can be accommodated in strongly bonding orbitals, A_{1g}, A_{2u}, E_{1u} and E_{1g} in ascending order of energy, two more in a non-bonding A_{1g} state related to the d_{z^2} orbital and four in weakly bonding E_{2g} molecular orbitals derived from metal d_{xy} and $d_{x^2-y^2}$ orbitals. Thus, as would be expected purely from symmetry considerations, the $3d_{xy}$ and $3d_{yz}$ are the metal atomic orbitals which contribute most to metal–ring bonding.

The chemical reactions of ferrocene are largely those which arise from the aromatic properties of the C_5H_5 rings, and electrophilic attack on them has been the subject of extensive study. There is evidence that electrophiles interact first with the iron:

$$(C_5H_5)_2Fe + E^+ \rightleftharpoons [(C_5H_5)_2Fe{-}E]^+ \rightarrow [(C_5H_5)Fe(C_5H_5E H)]^+ \rightarrow (C_5H_4E)Fe(C_5H_5) + H^+$$

The great thermodynamic stability of bis(cyclopentadienyl)iron suggested that other π-bonded organometallic compounds of iron should be capable of syntheses. One such compound which has proved of particular interest is cyclobutadieneiron tricarbonyl which can be made as yellow crystals by treating $Fe_2(CO)_9$ with either *cis*- or *trans*-3, 4-dichlorocyclobutene :

$$C_4H_4Cl_2 \xrightarrow{Fe_2(CO)_9} (C_4H_4)Fe(CO)_3$$

As in ferrocene, there is an extensive chemistry with the electrophilic substitution reactions. In Friedel–Crafts acylation, for example, the compound is much more reactive than benzene and almost as reactive as ferrocene itself.

Compounds derived from iron carbonyls and olefins are numerous. In 1930 a compound $C_4H_6Fe(CO)_3$ was made by the action of butadiene on $Fe(CO)_5$, but not until 1960 was it proved to be a π-complex. More recently it has been found that protonation of this compound gives rise to a π-allyl complex:

CH — CH
CH₂ CH₂ —H⁺→ CH₂ CH CH — CH₃
Fe Fe⁺
CO CO CO CO CO CO

which, in the form of its iodide, is the starting material for yet another extensive area of organometallic chemistry.

The α-bonded organometallic compounds of iron are far fewer than the π-complexes. The alkyls and aryls themselves are not stable enough to exist, but many compounds are known containing π-bonding ligands such as CO, Ph_3P, or $\pi\text{–}C_5H_5$ in addition to an Fe–C α-bond. An example is $\pi\text{–}(C_5H_5)\text{–}Fe(CO)_2CF_2CF_2H$, made by the action of perfluoroethylene on cyclopentadienyliron dicarbonyl hydride.

The organometallic chemistry of cobalt is rather similar to that of iron. Purple, air-sensitive crystals of cobaltocene can be made by treating $Co(CNS)_2$ with C_5H_5Na in liquid NH_3. Unlike ferrocene, it is paramagnetic ($\mu = 1.76\ \mu_B$) and is readily oxidised to diamagnetic $(C_5H_5)_2Co^+$ salts. Olefin, allyl and acetylene complexes of cobalt have also been made. Among the α-bonded organometallic compounds of cobalt the alkyl cobalt carbonyls are important in synthetic organic chemistry. The most important reaction is the reversible carbonylation:

$$RCo(CO)_4 + CO \rightleftharpoons RCOCo(CO)_4$$

The acyls so formed react with alcohols at about 320 K to give esters; thus alkyl halides can be converted to esters by reactions formulated:

$$RX + CO + Co(CO)_4^- \rightarrow RCOCo(CO)_4 + X^-$$
$$RCOCo(CO)_4 + R'OH \rightarrow RCOOR' + HCo(CO)_4$$

Although the catalytic effect of $Ni(CO)_4$ and its phosphine derivatives on the polymerisation of olefin has been studied in depth, few olefin complexes of nickel have been isolated. Perhaps the most important group of π-bonded organonickel compounds is that of π allyls. Bis (π-allyl) nickel itself is made as yellow, pyrophoric crystals by the action of allyl magnesium bromide on $NiBr_2$ in ether at 263 K. Its sandwich structure has been confirmed by X-ray analysis:

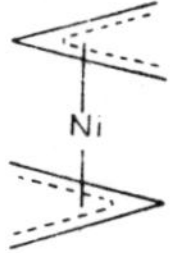

It is an active catalyst for the cyclotrimerisation of butadiene to cyclododeca-1,5,9-triene; π cyclopentadiene and π-cyclopentadiene compounds of nickel are also known. Emerald green nickelocene, $(\pi\text{–}C_5H_5)_2Ni$, is best made by addition of a solution of $NiCl_2$ in dimethyl sulphoxide to C_5H_5K in diglyme.

Of the α-bonded organonickel compounds, those made by the action of aryl magnesium halides on square $(R_3P)_2NiX_2$ complexes (X = halogen) are well established. These are yellow, diamagnetic, compounds of general formula $(R_3P)_2NiX(Ar)$ and $(R_3P)_2NiAr_2$ which have *trans*-square configurations.

COMPLEXES

Although some of the complexes of these metals have already been discussed their chemistry is so extensive that other aspects are worth attention.

BIPOSITIVE STATES

Iron (II)

Most of the complexes of iron(II) (d^6) are octahedral; but diamagnetic compounds are far less common than among the isoelectronic Co^{III} compounds, not unnaturally in view of the smaller charge on the iron. The six-co-ordinate, diamagnetic complexes include the hexacyanoferrates(II) and the tris (*ortho*-phenanthroline)iron(II) ions. The octahedral $[Fe(phen)_3]^{2+}$ ion (Fig. 6.6) is blood-red ; it is oxidised to pale blue $[Fe(phen)_3]^{3+}$ without any structural change. E° for the system = 1.14 V , making the compound also known as ferroin, a most useful redox indicator for the oxidation of

Fe^{2+} ion E°, Fe^{2+}/Fe^{2+} = 0.77 V) by cerium (IV) ion E°, Ce^{4+}/Ce^{3+} = 1.45 V.

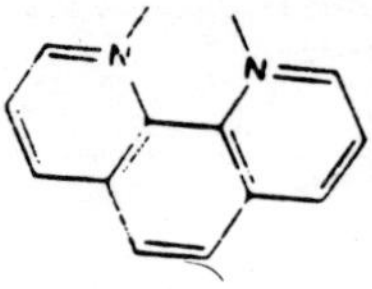

Fig. 6.6. The (phen) group in octahedral $[Fe(phen)_3]^{2+}$ ion.

The pentacyanonitrosylferrates, (the 'nitroprussides') are of interest. Cyano complexes in general do not easily form mixed complexes by replacement of CN^- groups and, in the hexacyanocomplexes of iron, only one CN^- can be replaced by NH_3, H_2O, CO, NO_2 (nitro) or NO (nitroso). Acidification of a $K_4Fe(CN)_6$–KNO_2 mixture gives first the pentacyanonitroferrate(II) ion, $[Fe(CN)_5NO_2]^{4-}$:

$$Fe(CN)_6{}^{4-} + NO_2{}^- = [Fe(CN)_5NO_2]^{4-} + CN^-$$

and then the pentacyanonitrosylferrate(II) ion, the 'nitroprusside' ion:

$$[Fe(CN)_5NO_2]^{4-} + 2\,H_3O^+ = [Fe(CN)_5NO]^{2-} + 3H_2O$$

The red sodium salt, $Na_2[Fe(CN)_5NO] \cdot 2\,H_2O$ is diamagnetic; surprisingly since the NO group is an odd-electron group which must confer paramagnetism should it be co-ordinated in the normal way. If the group is considered as co-ordinating in the form NO^+, however, not only is the diamagnetism comprehensible but the charge on the iron becomes clear: it is, in fact, +2.

The SH^- ion in alkaline solution converts the nitroprusside ion to purple $[Fe(CN)_5NOS]^{4-}$:

$$[Fe(CN)_5NO]^{2-} + OH^- + SH^- \rightarrow [Fe(CN)_5NOS]^{4-} + H_2O$$

High-spin octahedral iron(II) complexes are common. The pale-green hexa-aquoiron(II) ion, $Fe(H_2O)_6{}^{2+}$, present in dilute aqueous solution, owes its colour to a single absorption band with a peak at about 10,000 cm^{-1} (120 kJ mol^{-1}) in the i.r. which spreads into the visible spectrum. A few tetrahedral iron(II) complexes are known;

examples are the FeX_4^{2-} (X = halogen) anions already discussed, and neutral complexes such as $(Ph_3PO)_2FeX_2$. There are also some low-spin, square pyramidal complexes like the $[FeClO_4(OAsMe_3)_4]^+$ ion and a few square complexes, e.g. $Fe(C_6Cl_5)_2(PPhEt_2)_2$.

Cabalt (II)

The Co^{2+} ion is the only d^7 ion of common occurrence. The complexes formed from it are of several types; octahedral and tetrahedral are most common but square and 5-co-ordinate ones also occur. For a d^7 ion the ligand-field stabilisation should favour the octahedral configuration only slightly relative to the tetrahedral symmetry proves to be more common in Co^{II} complexes than in those of any other transition metal. Tetrahedral complexes of cobalt(II) with halide ions have already been mentioned, and there are others with monodentate pseudohalide ligands such as SCN^- Bidentate β-diketonate ions with large alkylgroups also give rise to tetrahedral co-ordination, but with small alkyl groups, as in acetylacetonate, an octahedral arrangement is favoured. With several singly-charged bidentate ions such as dimethylglyoximate and dithioacetylacetonate a square configuration is observed.

With strong-field ligands the tendency for Co^{II} complexes to oxidise to Co^{III} is very strong. Though low-spin octahedral cobalt(II) complexes such as $Co(diars)_3^{2+}$ have magnetic moments very close to spin-only values, the high-spin ones have large moments of 4.7 to 5.2 μB owing to large orbital contributions. The square complexes are all low-spin ones with magnetic moments in the range 2.2 to 2.7 μB.

Nickel (II)

Nickel in its dipositive state forms complexes of five structural types, octahedral trigonal bipyramidal, square pyramidal, tetrahedral and square. The aquo ligands in the octahedral $Ni(H_2O)_6^{2+}$ ion can be replaced by ammines to give ions such as $Ni(NH_3)_6^{2+}$ and $Nien_3^{2+}$. Their magnetic moments are in the range 2.9 to 3.4 μB, indicating that there are two unpaired electrons and that the orbital contribution to paramagnetism is variable but sometimes large. The absorption spectrum contains four bands in the visible and near ultraviolet, with ϵ_{max} less than 1.0 m^2 mol^{-1}

Hydrated nickel salts are apple-green, but the octahedral complexes with nitrogen ligands are usually blue or purple because the peaks shift towards the u.v. in the stronger ligand field :

$Ni(H_2O)_6^{2+}$		$Ni(NH_3)_6^{2+}$	
Wave number/cm^{-1}	$\in$ / m2 $^{mol-1}$	*Wave number*/cm^{-1}	$\in$ /m^2 mol^{-1}
8500	0.20	10750	0.40
13500	0.18	15150	0.50
15400	0.15	17500	0.48
25300	0.52	28200	0.63

It should be noted that where there are several absorption peaks a general shift to shorter wavelength does not necessarily mean a colour shift towards the red as in the simple d^1 case. In the foregoing examples the ammine complex is deep blue because the principal absorption peak is in the ultraviolet whereas the aquo complex is green because there is strong absorption in the violet.

An important class of complexes incorporating 5-co-ordinate nickel(II) is exemplified by the ion $NiN[CH_2CH_2N(CH_3)_2]_3Br^+$:

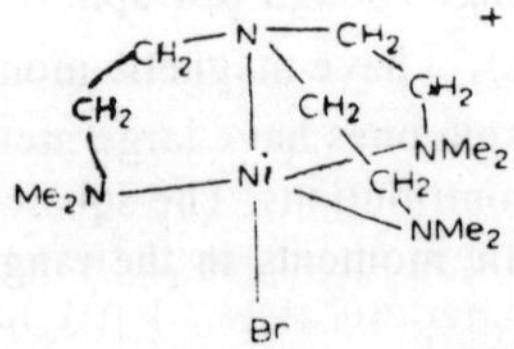

Among other quadridentate 'tripod' ligands which give rise to this symmetry are $N(CH_2CH_2PPh_2)_3$ and $P(o\text{–}C_6H_4SMe)_3$. However the ion $Ni(CN)_5^{3-}$, which contains only unidentate ligands, is also known to exist in a trigonal bipyramidal form.

Several of the paramagnetic nickel(II) complexes, once thought to be tetrahedral on stoichiometric grounds, have been found to be associated and octahedral. Nevertheless, increasing numbers of authentic tetrahedral complexes are being discovered: examples are $NiI_2(PPh_3)_2$, $NiI_2(py)_2$ and the chelate isopropyl- and sec-butylsalicylaldimine complexes. They are usually blue compounds and their absorption in the red part of the spectrum (~ 15000 cm^{-1}) is fairly strong ($\in$~ 16m^2 mol^{-1}).

Typical square planar nickel(II) complexes are uncharged complexes:

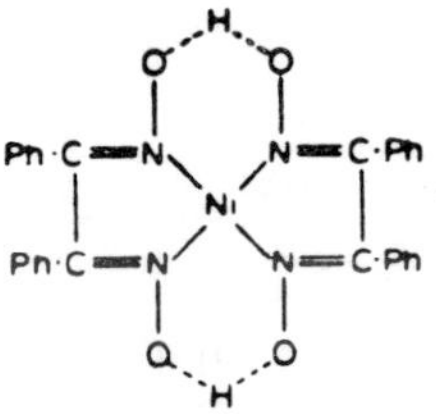

Bis(diphenylglyoxime)nickel(II)

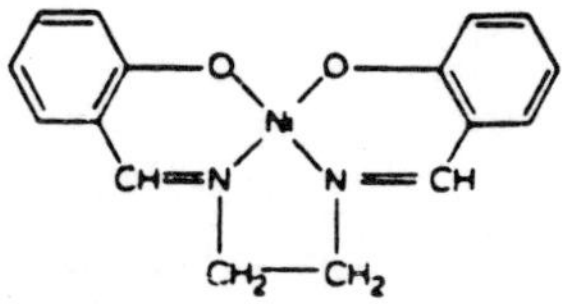

Salicylaldehydo-ethylenediaminenickel(II)

The compounds are usually pink : there are at most three bands in the visible absorption spectrum, the strongest being at about 22,000 cm^{-1} ($\in$ ~ 15 to 35 m^2 mol^{-1}).

Many square planar Ni^{II} complexes which are diamagnetic in the solid state become paramagnetic in solution. Bis(*N*-methylsalicylaldimine)nickel(II) shows weak absorption bands which indicate that some of the nickel atoms are in a triplet ground state even in the diamagnetic solid. When the solid is dissolved in chloroform these bands become stronger, and the compound becomes paramagnetic (μ = 2.3 μB. The solvent weakness the ligand field and uncoupling of spin occurs. In pyridine the same compound has the type of absorption spectrum characteristic of a paramagnetic octahedral Ni^{II} complex and the magnetic moment reaches 3.1 μB. Molecules of the base can enter the fifth and sixth co-ordination positions, thereby raising the energy of the d^{z2} orbital and causing the ligands in the *xy* plane to be repelled :

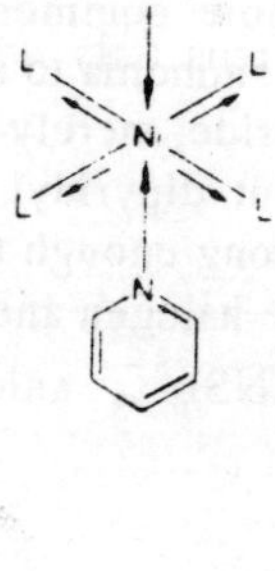

Another way in which planar nickel(II) complexes are converted into tetragonal or octahedral ones is by association either in solution or in the crystal.

There is evidence for equilibria between diamagnetic planner and paramagnetic tetrahedral isomers in solution and melts. The phenomenon is shown by a number of bis(*N*-sec-alkysalicylaldimine) nickel(II) complexes:

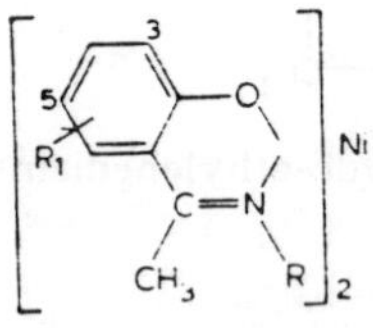

R = sec. alkyl

R_1 = alkyl at 3 or 5 position

Of the solids, some are diamagnetic and some paramagnetic (μ ∈ 3.3 μB).

When dissolved in toluene the diamagnetic ones become paramagnetic, the paramagnetic ones less so.

The existence of equilibria between conformational isomers:

tetrahedral (paramagnetic) ⇌ planner (diamagnetic)

is evident from the absorption spectra.

TERPOSITIVE STATES

Iron (III)

Most Iron(III) complexes are octahedral, but tetrahedral anions such as $FeCl_4^-$ are known and there is the pentagonal bipyramidal $[Fe(edta)H_2O]^-$. Of the octahedral complexes, those with ligands such as oxalate, phosphate and β-diketones, which co-ordinate through oxygen atoms, are much more common than those formed from nitrogen donors. Addition of ammonia to an iron(III) salt, even in the presence of ammonium chloride, merely causes the precipitation of hydrated iron(III) oxide. But dipyridyl and *ortho*-phenanthroline, which exert ligand fields strong enough to cause spin-pairing, form stable complexes. Among the halogen and pseudohalogen complexes are FeF_6^{3-}, $Fe(CN)_6^{3-}$, $Fe(CNS)_6^{3-}$; anions containing five of one

type of ligand and one of another are rather common; $[Fe(CN)_5H_2O]^{2-}$; and $FeF_5(H_2O)]^{2-}$; are examples.

Iron(III) (d^5) compounds might be expected to have very similar absorption spectra to those of the isoelectronic Mn^{II} , but the weak d–d transitions are in fact masked by the spread into the visible spectrum of strong charge-transfer peaks belonging to the near ultraviolet.

Binuclear, oxygen-bridged iron(III) compounds are of interest. They are of two kinds, the dihydroxo-bridged complexes exemplified by the $[Fe(H_2O)_4(OH)_2Fe(H_2O)_4]^{4+}$ ion which exists in aqueous solutions of Fe^{3+} in the pH range 2–3: and the monoxo-bridged

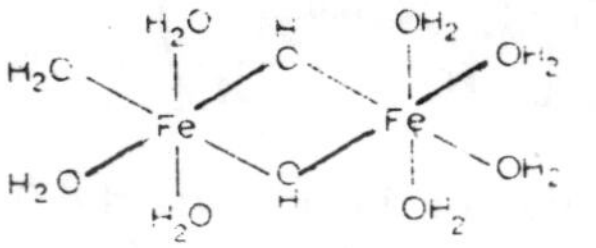

complexes exemplified by the $[Fe(Hedta)]_2O^{2-}$ ion in which the two iron atoms are bound to a single oxygen:

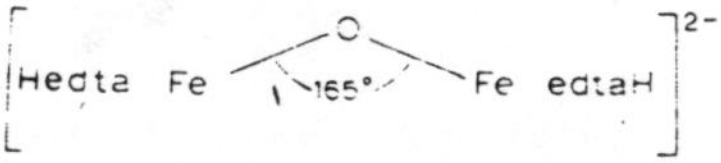

Cobalt (III)

There is a wide range of cobalt(III) complexes with nitrogen donors, and most of the pioneer investigation of transition-metal complexes was done on cobaltammines. Although cobalt(III) salts are oxidised only with extreme difficulty ($E°$, Co^{3+}/Co^{2+} = + 1.84 V), H_2O_2 or even air will oxidise Co^{II} to Co^{III} in the presence of cyanide or ammonia $E°$, $Co(CN)_6^{3-}/Co(CN)_6^{4-}$ = –0.83 V. Hexa-amminecobalt (III) chloride, $Co(NH_3)_6Cl_3$, is readily made by oxidising a solution of $CoCl_3$, NH_3 and NH_4Cl containing active charcoal as a catalyst :

$$2\,CoCl_2 + 2\,NH_4Cl + 10\,NH_3 + H_2O_2 \rightarrow 2\,Co(NH_3)_6Cl_3 + 2\,H_2O$$

This compound contains the $Co(NH_3)_6^{3+}$ ion. Treatment of $Co(NO_3)_2$, NH_4NO_3 and NH_3 with H_2O_2 in the absence of charcoal gives a nitratopenta-amminecobalt nitrate however, $[CoNO_3(NH_3)_5](NO_3)_2$. Other series can be made by suitable methods.

The compound $K_6Co_2(CN)_{10}$ is easily oxidised in weak acid to the very stable $K_3Co(CN)_6$. The so-called cobaltinitrites are also Co^{III} complexes. Addition of aqueous KNO_2 to a Co^{2+} solution containing acetic acid gives a yellow precipitate of potassium hexanitrocobaltate(III), $K_3Co(NO_2)_6$.

$$Co^{2+} + 7\,NO_2^- + 2H^+ \rightarrow Co(NO_2)_6^{3-} + NO + H_2O$$

When the violet salt, chloroaquotetra-amminecobalt(III) sulphate, $[Co(NH_3)_4(H_2O)Cl]SO_4$, is treated with cold alkali the asymmetric ion

$$\left(Co\left[\left\langle\begin{matrix}H\\O\\ \\O\\H\end{matrix}\right\rangle Co(NH_3)_4\right]_3\right)^{6+}$$

is produced. The chloride of this ion was the first purely inorganic (non-carbon) compound to be resolved (Werner, 1914).

Another type of polynuclear cobalt complex is produced by aerial oxidation of ammoniacal Co(N Q)$_2$ solution. This is the μ-peroxo-bis {penta-amminecobalt(III)} ion, $(NH_3)_5Co{-}O{-}O{-}Co\,(NH_3)_5]^{4+}$. Further

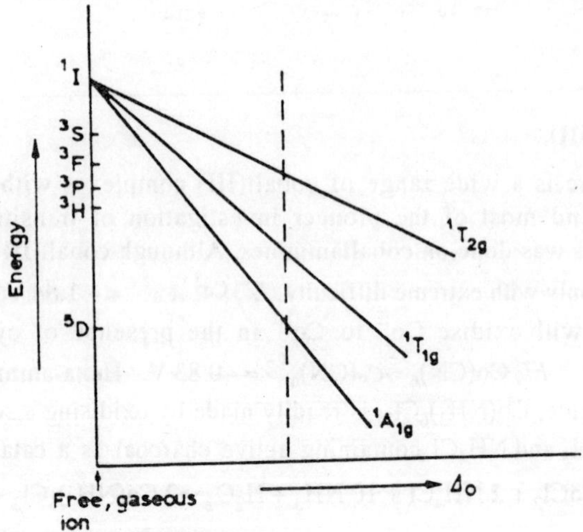

Fig. 6.7. Energy level diagram for d^6 ion in an octahedral field (singlet states only).

oxidation gives the quinquepositive ion $[(NH_3)_5CoO_2Co(NH_3)_5]^{5+}$. It was shown by electron-spin resonance, to contain cobalt atoms of identical oxidation number, not Co^{III} and Co^{IV} as previously thought. X-ray studies indicated the axis of O–O to be perpendicular to the Co–Co axis.

All cobalt (III) complexes except CoF_6^{3-} have singlet ground states and are diamagnetic. The energy level diagram for the d^6 ion in the octahedral field, with only the singlet states shown, has the form illustrated in Fig. 6.7.

There are usually two absorption bands in the visible spectrum. For symmetrical complexes, CoX_6, each of these bands, when plotted on a wavenumber scale, is almost perfectly symmetrical about its peak. But for less symmetrical complexes such as CoX_4Y_2 there is considerable splitting of the $^1T_{1g}$ state. Furthermore, the splitting is greater for the *trans* form than for the *cis* form of such a complex. For the *trans* the absorption peak at the lower energy ($^1A_1g \rightarrow {}^1T_1g$ transition is split into two peaks, for the *cis* the peak differs from that of the symmetrical CoX_6 in having a 'shoulder'. Moreover because a *cis* octahedral isomer is without a centre of symmetry the intensity of absorption is stronger. Typical absorption curves for *cis* and *trans* isomers of the CoX_4Y_2 type are illustrated in Fig. 6.8.

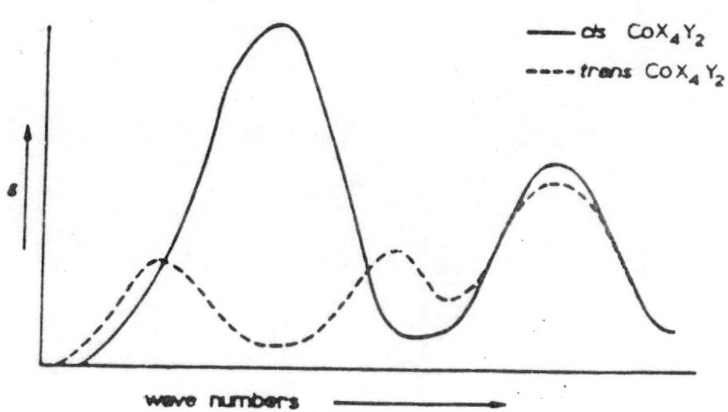

Fig. 6.8 Typical absorption curves for cis and trans $C_oX_4Y_2$ isomers

Nickel (III)

This is an unusual oxidation state in nickel, but a few complexes with the element in this state are known. The trimethylphosphine complexes of nickel(II) halides, $NiX_2(PMe_3)_2$, are oxidised to nickel (III) complexes even by atmospheric oxygen. When the appropriate nitrosyl halide is used as the oxidising agent the compounds formed are $NiX_3(PMe_3)_2$; they are monomeric in solution.

Higher oxidation states

Treatment of alcoholic $FeCl_3$ with o–$C_6H_4(AsMe_2)_2$ (diarsine) gives a brilliant red precipitate of $[Fe^{III}Cl_2(diarsine)_2]FeCl_4$ which, when warmed in nitrobenzene with concentrated HNO_3, gives a compound $[Fe^{IV}Cl_2(diarsine)_2](FeCl_4)_2$. The magnetic moment of 2.98 μB is consistent with the presence of two unpaired electrons in a tetragonally distorted set of t_{2g} orbitals. Other compounds of Fe^{IV}, Co^{IV} and Ni^{IV} are principally fluoroanions which have already been discussed, but some diamagnetic Ni^{IV} heteropoly-anions such as $[NiMo_9O_{32}]^{6-}$ and $[NiNb_{12}O_{38}]^{12-}$ contain nickel in this formal oxidation state.

The charge number +6 does not occur in Co and Ni, and in Fe only as the ferrates such as $BaFeO_4$ which have already been discussed.

CHAPTER 7

The Platinum Metals

INTRODUCTION

The second triad of Group VIII, ruthenium, rhodium and palladium, and the third triad, osmium, iridium and platinum, are sufficiently alike in physical and chemical character to be known collectively as the platinum metals. Vertical similarities are much stronger between members of these triads than between them and

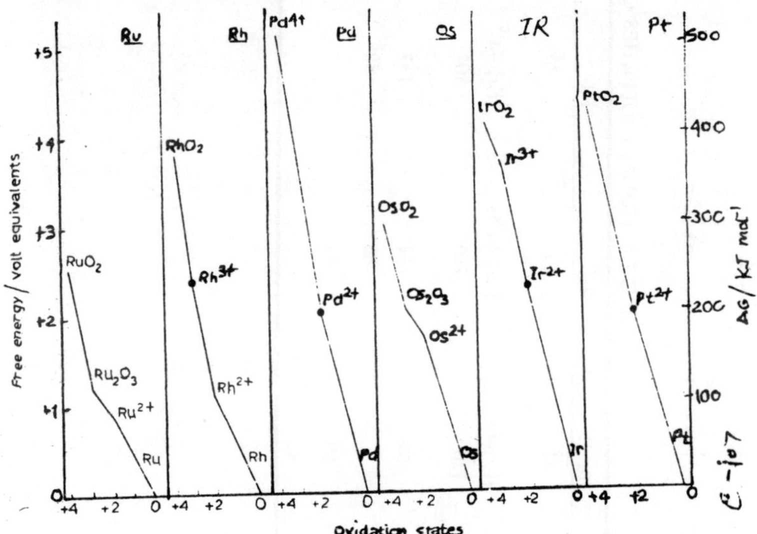

Fig. 7.1. Free energies of oxidation states of the platinum family, relative to the metal, in aqueous solution at pH = 0.

TABLE 7.1 ATOMIC PROPERTIES OF Ru, Rh, Pd, Os, Ir and Pt

	Ru	Rh	Pd	Os	Ir	Pt
Z	44	45	46	76	77	78
Electron configuration	[Kr] $4d^7 5s^1$	[Kr] $4d^8 5s^1$	[Kr] $4d^{10}$	[Xe] $4f^{14} 5d^6 6s^2$	[Xe] $4f^{14} 5d^7 6s^2$	[Xe] $4f^{14} 5d^9 6s^1$
$I(1)$/kJ mol^{-1}	720	720	804	840	840	870
Metallic radius/pm	134	134	137	135	136	138
$r_{M^{2+}}$/pm		86	80			80
$r_{M^{3+}}$/pm	69	69				
$r_{M^{4+}}$/pm	65			67	66	

iron, cobalt and nickel, but some correlations can be observed. Thus Fe, Ru and Os form monomeric carbonyls, $M(CO)_5$, the ions MO_4^{2-} and the very stable complex ions $M(CN)_6^{4-}$, $M(phen)_3^{2+}$ and $M(bipy)_3^{2+}$. Co, Rh and Ir all form the dimeric carbonyls $M_2(CO)_8$ and the ions $M(CN)_6^{3-}$ and $M(C_2O_4)_3^{3-}$, and Ni, Pd and Pt have the very stable $M(CN)_4^{2-}$ ions in common as well as neutral complexes with oximes, the best known being the dimethylgloxime complexes.

Metallic radii in the platinum metals are almost uniform (Table 7.1) ; not surprisingly the elements show great physical similarities. The first ionisation energies are not much greater than those of the d-block elements which precede them but the electrode potentials are strongly positive, a major factor being undoubtedly the very high sublimation energies.

STABILITY OF OXIDATION STATES IN AQUEOUS SOLUTION

The free energies of oxidation up to +4, relative to the metal, are illustrated in Fig. 7.1. Notable features of the diagram are the relative stability of the +3 state in the elements Ru and Os, the tendency of this state to disproportionation in Rh and Ir (cf. their congeners), and the absence of the +3 state in Pd ant Pt (cf. Ni).

Higher oxidation states of Ru and Os exist in solution, for instance in RuO_4^- and $OsO_4(OH)_2^-$, but they are unstable at low pH and their redox potentials are in doubt.

THE ELEMENTS

Reparation And Properties

The six metals together comprise about $2 \times 10^{-6}\%$ of the lithosphere and are often found native. Osmiridium is a natural alloy of osmium and iridium.

Important ores are sperrylite, $PtAs_2$, cooperite, PtS, and braggite, (Pt, Pd, Ni)S. The Sudbury (Canada) nickel and copper sulphide deposits contain about 2 P.P.m. of the platinum metals, other than osmium. The metals along with copper and gold, are segregated in the undissolved sludges from the electro-refining of Ni and Cu and in the involatile residues after the removal of $Ni(CO)_4$ in the carbonyl process. The Sudbury seperation (Fig. 7.2) begins with an extraction by hot aqua regia in which only platinum, paladium and gold dissolve ; the various metals are recovered as indicated as Fig. 7.2.

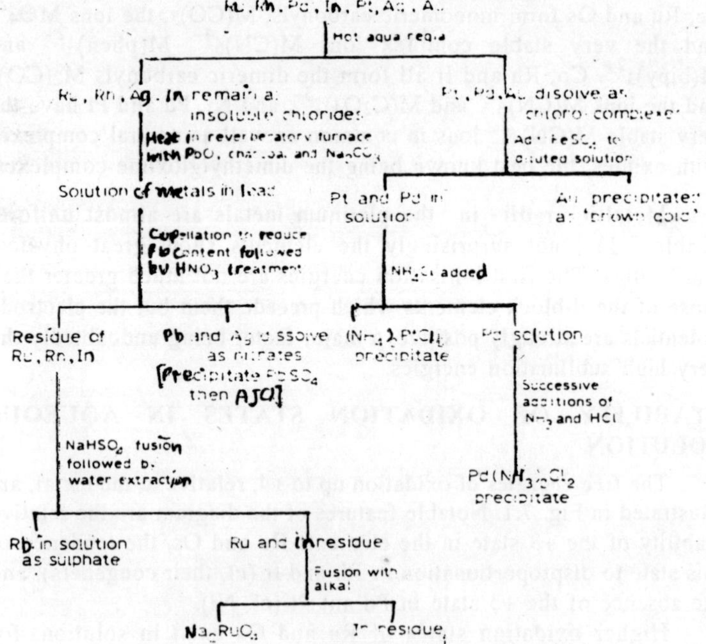

Fig. 7.2. Separation of the platinum metals.

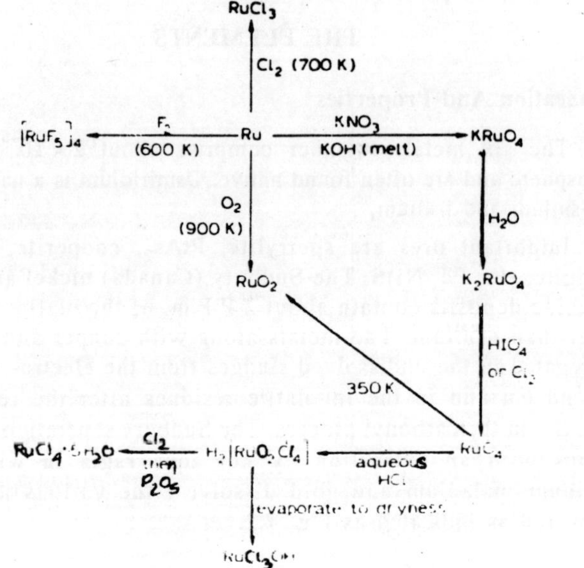

Fig. 7.3. Ruthenium, some reactions and preparations.

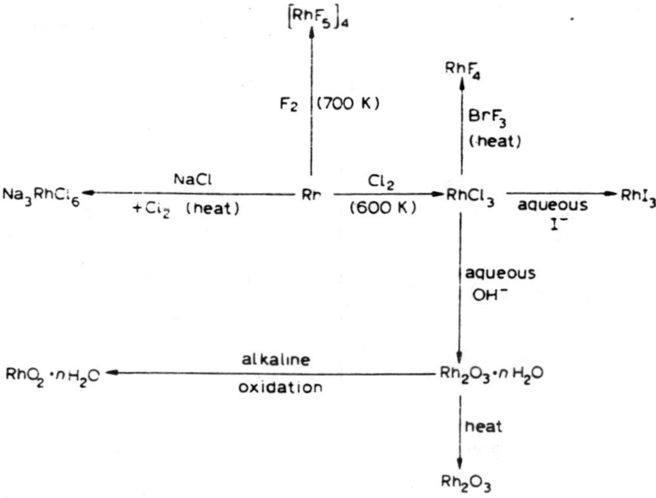

Fig. 7.4. Rhodium, some reactions and preparations.

Total world production of platinum metals exceeds 100 tons per annum.

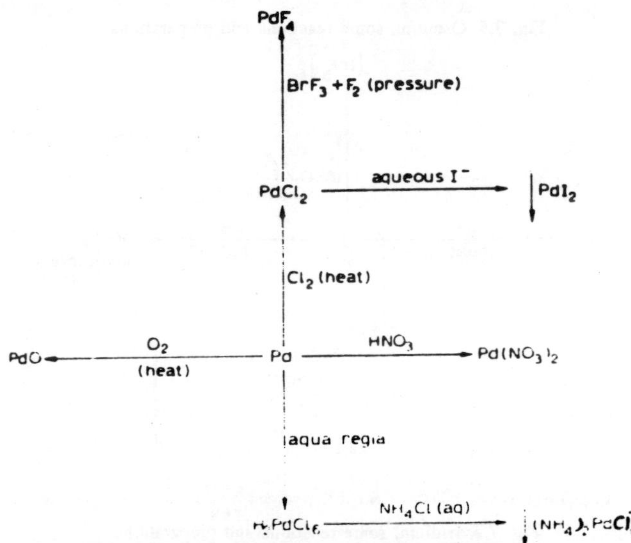

Fig. 7.5. Palladium, some reactions and preparations.

The densities of the metals fall into two groups, those in the third triad being the highest of all the elements. Ru and Os have h.c.p. structure, the rest are cubic close-packed. Melting points, hardness

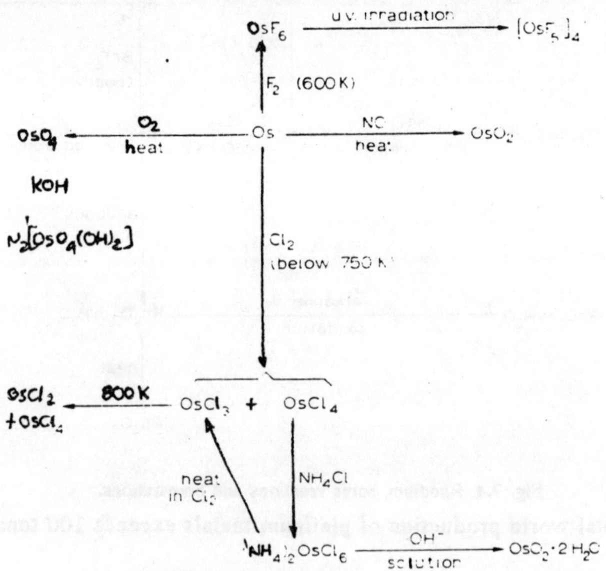

Fig. 7.6. Osmium, some reactions and preparations.

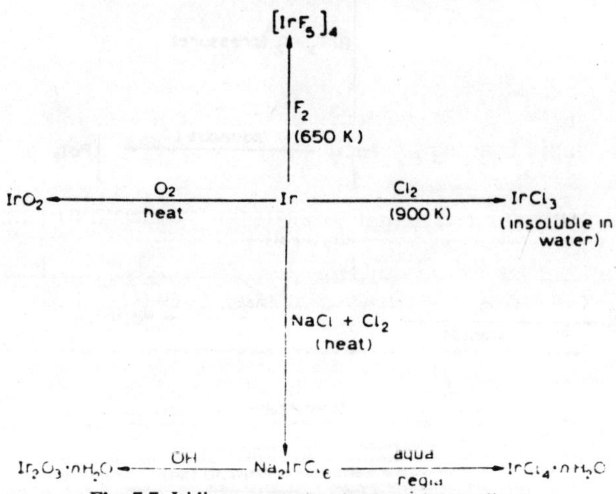

Fig. 7.7. Iridium, some reactions and preparations.

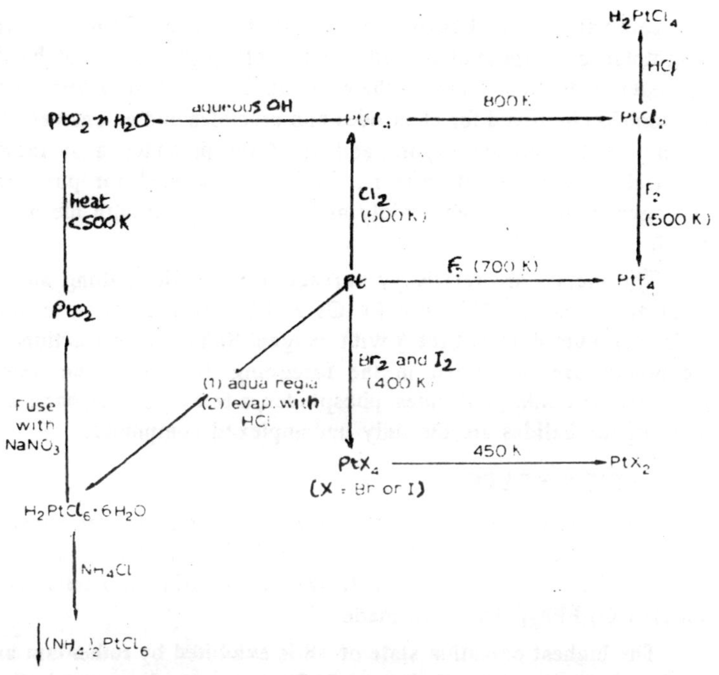

Fig. 7.8. Platinum, some reactions and preparations.

and mechanical strength decrease progressively in each period. These differences have been attributed to the decreasing availability of electrons for bonding among the atoms in the solid state, since electrons which become paired in atomic d orbitals fail to participate in Metallic bonding.

TABLE 7.2 PHYSICAL PROPERTIES OF THE PLATINUM METALS

	Ru	Rh	Pd	Os	Ir	Pt
M.p./K	2720	2240	1830	2970	2730	2040
ρ/g cm^{-3}	12.45	12.41	12.02	22.61	22.65	21.45

Several of the metals are useful in catalysis. Platinum is used as surface catalyst in (a) the hydrogenation of olefins and acetylenes, (b) the oxidation of NH_3, to HNO_3, (c) the reforming of hydrocarbons and (d) the oxidation of hydrocarbons in fuel cells. Rhodium supported in a finely divided state on alumina is particularly useful for catalysing the hydrogenation of benzene. Some of the intermetallic compounds of iridium, notably Ti_3Ir, $ZrIr_2$ and Nb_3Ir, are superconductors. Many of the elements are resistant to corrosion : platinum is employed for electrodes and crucibles and extensively for spinnerets to extrude rayon, and rhodium provides a surfacing material or searchlight mirrors. Palladium is used for purifying hydrogen which diffuses rapidly and selectively through the warm metal.

The metals are relatively unreactive. Palladium along among them dissolves in HNO_3 and Pt, Os and Pd only dissolve in aqua regia. Platinum does not react with oxygen. Some of the reactions of the metals are illustrated in the foregoing diagrams. The oxide, sulphides, selenides, tellurides, phosphides, a few of the sulphates and some of the halides are the only uncomplexed compounds.

OXIDATION STATES

Only a few examples of negative oxidation states are known : rhodium forms the carbonylate anion $Rh(CO)_4^-$, and there is some evidence for the $Ru(CO)_4^{2-}$ ion. In the case of iridium a tetrahedral ion $[Ir(CO)_3PPh_3]^-$ has been made.

The highest oxidation state of +8 is exhibited by ruthenium and osmium in their oxides, RuO_4 and OsO_4, but in rhodium and iridium the maximum charge number is only +6. Indeed the general charge number pattern of the preceding transition elements changes at the Periodic Table ; thereafter the elements tend show less diversity of oxidation state. This is evinced by the closer similarity between Ni, Pd and Pt than between the respective members of the two preceding vertical groupings.

The charge numbers +1 and +5 are uncommon. All the elements except palladium form hexafluorides, but in palladium the highest known oxidation state is +4.

HALIDES

The hexafluorides are known for all the platinum metals except palladium. They are volatile, reactive substances which are normally kept in nickel or monel apparatus because they attack glass, some of them even at room temperatures. The compounds are normally made by direct combination of the elements followed by rapid condensation on a cold finger. Dark-red crystals of PtF_6 so made have extraordinary oxidising powers. The vapour was discovered to oxidise O_2 to O_2^+, a result which stimulated the work which led to the preparation of the first compound of cationic xenon, $XePtF_6$. The hexafluorides have octahedral molecules.

The five metals which form hexafluorides also form tetrameric compounds $(MF_5)_4$ in which the octahedrally co-ordinated metal atoms are linked by M—F—M bridges similar to those in the corresponding compounds of Nb, Ta and Mo. They are normally obtained by direct fluorination at temperatures rather higher than those used for the preparation of hexafluorides.

All six metals form tetrahalides. Brick-red PdF_4 is made by fluorination of a compound formerly called PdF_3 but now known to be $Pd^{11}(Pd^{IV}F_6)$ which is made by treating $PdBr_2$ with BrF_3 and heating to 450 K the adduct Pd_2F_6. $2BrF_3$ so formed. RhF_4 and PtF_4 can also be made from BrF_3 adducts :

$$RhCl_3 \xrightarrow{BrF_3} RhF_4 . 2\,BrF_3 \xrightarrow{heat} RhF_4$$

$$Pt \xrightarrow{BrF_3} PtF_4 . \quad 2\,BrF_3 \xrightarrow{heat} PtF_4$$

Red-brown $PtCl_4$ is obtained by heating to 570 K the compound $H_2PtCl_6 . 6\,H_2O$ obtained by dissolving Pt in aqua regia. The other two tetrahalides of platinum are dark-brown compounds made by direct combination of the elements.

Of the trifluorides, RuF_3 is best made by reducing RuF_5 with iodine at 520 K :

$$5\,RuF_5 + I_2 \rightarrow 5\,RuF_3 + 2\,IF_5$$

and IrF_3 by reducing IrF_6 with Ir. Most of the other trihalides shown in Table 37.4 are made by direct combination of the elements or by precipitation from solution :

$$RhCl_3 + 3\,I^- \rightarrow RhI_3 + 3\,Cl^-$$

The dark-red $PdCl_2$ and $PdBr_2$ are made from the elements at red heat, but PdF_2 is best made by refluxing $Pd^{II}[Pd^{IV}Cl_6]$ with SeF_4 :

$$Pd_2Cl_6 + SeF_4 \rightarrow 2\,PdF_2 + SeCl_6$$

Dark-red $PtCl_2$ can be obtained by thermal decompositions of $PtCl_4$, but the dibromide and di-iodide are so obtained only with difficulty.

TABLE 7.3 COMPOUNDS AND IONS REPRESENTATIVE OF THE OXIDATION OF Ru, Rh, Pd, Os, Ir AND Pt

Oxidation number	Ru	Rh	Pd	Os	Ir	Pt
–1		$Rh(CO)_4$			$[Ir(CO)_3PPh_3]^-$	
0	$Ru(CO)_5$	$Rh_2(CO)_8$	$Pd(NCR)_2$	$Os(CO)_5$	$Ir_4(CO)_{12}$	$Pt(PR_3)_4$
+1		$[RhCl(CO)_2]_2$		$Os(NH_3)_6BR$	$IrClCO(PEt_3)_2$	
+2	$Ru(NH_3)_6^{2+}$	$RhCl(dipy)_2^+$	PdO	$Os(CN)_6)^{4-}$	$Ir(NH_3)_4Cl_2$	PtS
+3	$RuCl(NH_3)_5^{2+}$	$RhCl_6^{3-}$		$OsCl_6^{3-}$	$IrCl_6^{3-1}$	
+4	$RuCl_6^{2-}$	RhF_6^{2-}	$PdCl_6^{2-}$	$OsCl_6^{2-}$	IrO_2	$PtCl_6^{2-}$
+5	RuF_5			OsF_6^-	IrF_6^-	PtF_6^-
+6	RuF_6	RhF_6		OsF_6	IrF_6	PtF_6
+7	RuO_4^-			$OsOF_5$,		
+8	RuO_4			OsO_4		

TABLE 7.4 HALIDES OF THE PLATINUM METALS

Oxidation state	Ru	Rh	Pd	Os	Ir	Pt
+2	$RuBr_2$		PdF_2, $PdCl_2$ $PdBr_2$, PdI_2	OsI_2		$PtCl_2$ $PtBr_2$, PtI_2
+3	RuF_3, $RuCl_3$ $RuBr_3$, RuI_3	RhF_3, $RhCl_3$ $RhBr_3$, RhI_3		$OsCl_3$ $OsBr_3$ OsI_3	IrF_3, $IrCl_3$ $IrBr_3$, IrI_3	
+4	RuF_4	RhF_4	PdF_4	OsF_4, $OsCl_4$ $OsBr_4$		PtF_4, $PtCl_4$ $PtBr_4$, PtI_4
+5	$(RuF_5)_4$	$(RhF_5)_4$		$(OsF_5)_4$	$(IrF_5)_4$	$(PtF_5)_4$
6	RuF_6	RhF_6		OsF_6	IrF_6	PtF_6

HALOGEN COMPLEXES

The principal fluoro- and chloro-anions are shown in Table 7.5.

TABLE 7.5 PRINCIPAL FLUORO- AND CHLORO- ANIONS OF THE PLATINUM METALS

Charge number	Ru	Rh	Pd	Os	Ir	Pt
+2			$PdCl_4^{2-}$			$PtCl_4^{2-}$
+3	RuF_6^{3-}	RhF_6^{3-}			IrF_6^{3-}	
	$RuCl_6^{3-}$	$RhCl_6^{3-}$		$OsCl_6^{3-}$	$IrCl_6^{3-}$	
+4	RuF_6^{2-}	RhF_6^{2-}	PdF_6^{2-}	OsF_6^{2-}	IrF_6^{2-}	PtF_6^{2-}
	$RuCl_6^{2-}$	$RhCl_6^{2-}$	$PdCl_6^{2-}$	$OsCl_6^{2-}$	$IrCl_6^{2-}$	$PtCl_6^{2-}$
+5	RuF_6^-	RhF_6^-		OsF_6^-	IrF_6^-	PtF_6^-

The chloro-complexes can sometimes be made by heating the metal with an alkali-metal chloride in a stream of chlorine :

$$2\,Rh + 6\,NaCl + 3\,Cl_2 \rightarrow 2\,Na_3RhCl_6$$

Another common method of preparation is to add NH_4Cl or KCl to the complex chloro-acid, ammonium and potassium salts being sparingly soluble as a rule :

$$H_2PtCl_6 + 2\,NH_4Cl \rightarrow (NH_4)_2PtCl_6 + 2\,HCl$$

$$H_2OsCl_6 + 2\,NH_4Cl \rightarrow (NH_4)_2OsCl_6 + 2\,HCl$$

The ammonium salt is frequently used in purification processes as it gives the metal on heating. The addition of KCl to a Na_3RhCl_6 solution precipitates the 5-co-ordinate K_2RhCl_5.

Ruthenium (III) exists in $RuCl_4^-$, $RuCl_5^{2-}$ and $RuCl_7^{4-}$, ions of unknown structure, as well as in $RuCl_6^{3-}$.

Platinum resists attack by concentrated aqueous HCl except in the presence of KCl or RbCl which, by forming insoluble chloroplatinates, disturb the equilibrium which is normally in favour of the metal. Similarly, gaseous HCl reacts with Pt in the presence of KCl, K_2PtCl_4 is first formed and subsequently disporportionates to Pt and K_2PtCl_6, on cooling.

Of all the chloro-complexes, only the chloroplatinates(II) and (IV) are not hydrolysed in aqueous solution at pH 7. The rest give precipitates of hydrated oxides under these conditions.

Of the fluorides, K_3RuF_6 is made by fusing $RuCl_3$ with KHF_2, and K_3RhF_6 by fusing $K_3Rh(NO_2)_6$ with KHF_2. Fluorocomplexes of +5 states can be made by fluorination of mixtures of alkali-metal halides with platinum-group halides :

$$RuCl_3 + MCl + 3\ F_2 \rightarrow MRuF_6 + 2\ Cl_2$$

$$2\ OsCl_4 + 2\ MCl + 6\ F_2 \rightarrow 2\ MOsF_6 + 5\ Cl_2$$

Bromine trifluoride is also used as the fluorinating agent :

$$IrBr_34 + MCl \xrightarrow{BrF_3} MIrF_6$$

These +5 complexes are unstable in water or dilute aqueous alkali. Thus when $KIrF_6$ is treated with dilute KOH it is converted to K_2IrF_6, an Ir^{IV} complex, which remains unchanged by the water. The hexafluororuthenates and hexafluoro-osmates(IV) are made similarly.

These are bromocomplexes corresponding to some of the chlorocomplexes listed above ; among the most important are the salts of $IrBr_6^{2-}$, $OsBr_6^{2-}$ and $PtBr_4^{2-}$. Few iodocomplexes are known.

OXIDES

The principal oxides of the platinum metals are shown in Table 7.6. A number of other solids of uncertain composition appear to exist and the oxygen chemistry of the group needs further investigation. Black PdO can be made by direct combination of the elements or by fusing $PdCl_2$ with Na_2CO_3, leaching out the soluble sodium salts and then dehydrating. The oxide, which is always somewhat oxygen-deficient, has a tetragonal lattice (Fig. 7.9) in which the

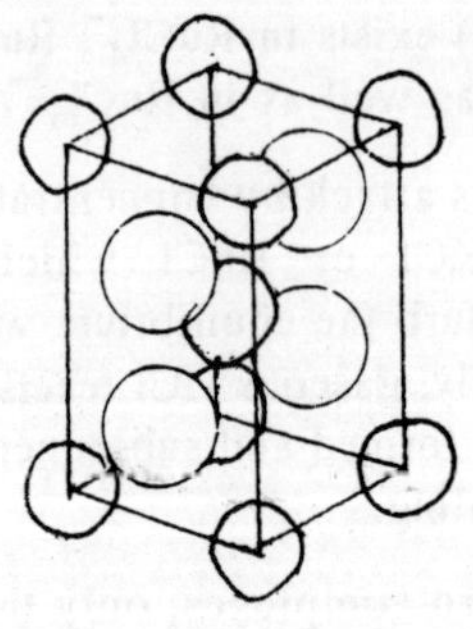

Fig. 7.9. Unit cell of PdO.

metal atoms have four coplanar bonds as in Pd^{II} complexes, and the oxygens are tetrahedrally co-ordinated.

A yellow, hydrated oxide $Rh_2O_3 . n\,H_2O$ is precipitated when a slight excess alkali is added to an $RhCl_3$ solution. The anhydrous oxide, which has the corundum structure, is best made from it by converting it with nitric acid to $Rh(NO_3)_3 . 2\,H_2O$ which is then decomposed by heating. If the solid so obtained is heated in oxygen under pressure it is converted to black RhO_2. A grey, hydrated iridium(III) oxide can be obtained by heating K_3IrCl_6 with Na_2CO_3 and removing the soluble products with water, but it is at least partly oxidised in air to a hydrated IrO_2. Black, anhydrous IrO_2 and black RuO_2 can both be made by heating the metals in oxygen, but OsO_2 is best made by the action of nitric oxide on the metal at 920 K. The foregoing dioxides all have the rutile structure. Platinum does not react directly with oxygen, being in this respect the most noble of the six metals. A brown oxide of approximate composition PtO_4 but unknown structure is formed by fusing H_2PtCl_6 and $NaNO_3$ at 750 K, washing free of sodium salts and then drying the product.

TABLE 7.6 ANHYDROUS OXIDES OF THE PLATINUM METALS

Charge number	Ru	Rh	Pd	Os	Ir	Pt
+2			PdO			
+3		Rh_2O_3				
+4	RuO_2	RhO_2		OsO_2	IrO_2	PtO2
+8	RuO_4			OsO_4		

The tetroxides of ruthenium and osmium are solids of low *m.p.* (RuO_4, 298 K and OsO_4, 314 K). Orange RuO_4 volatilises when a stream of chlorine is passed through an acidified solution of a ruthenate. It sublimes in a vacuum, but decomposes explosively into RuO_2 and oxygen at about 450 K. Colourless OsO_4 is much more thermally stable ; it can be made by heating the finely divided metal in oxygen. Both compounds contain tetrahedral molecules ; they are extremely soluble in CCl_4, which can be used to extract them from aqueous solutions. They are both strong oxidising agents and both can be considered to be acidic oxides. They differ in their behaviour towards alkalis, however. OsO_4 dissolves to give the ion $[OsO_4(OH)_2]^{2-}$ but RuO_4 liberates oxygen from the water and gives eventually a ruthenate(VI) ion :

$$2\,RuO_4 + 4\,OH^- \rightarrow 2\,RuO_4^{\,2-} + 2\,H_2O + O_2$$

SULPHIDES

Both palladium and platinum form monosulphides. PdS can be obtained as a grown precipitate by passing H_2S into a solution of $PdCl_4^{2-}$, and the grey PtS is made by heating together $PtCl_2$, Na_2CO_3 and sulphur. In both, the metal atom is surrounded by sulphur atoms in approximately square-planar co-ordination (Fig. 7.10). All the metals except palladium form disulphides which are obtained by direct combination of the elements. RuS_2, RhS_2 and OsS_2 have the cubic, pyrites structure containing S_2^{2-} ions and are therefore compounds of the metals in their +2 states, but PtS_2 has the CdI_2 structure. Iridium forms a trisulphide IrS_3 when $IrCl_3$ is heated with an excess of sulphur in a sealed tube, and rhodium forms a compound Rh_2S_5 in a similar reaction. The platinum metals also react with selenium and tellurium, the principal products being diselenides and ditellurides. In general they resemble the sulphides structurally, thus the compounds of Ru, Rh and Os have the pyrites structure whereas PtSe2 and $PtTe_2$ have the CdI_2 lattice.

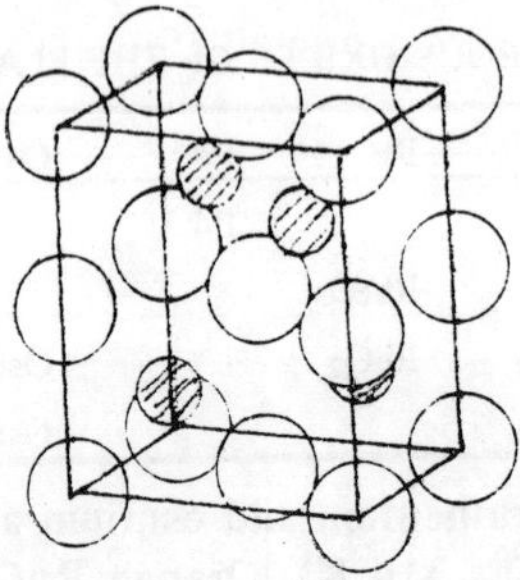

Fig. 7.10. Unit cell of PtS.

OXOACID SALTS

Various yellow hydrates of $Rh_2(SO_4)_3$ are known. The red sulphate $Rh_2(SO_4)_3 \cdot 6\ H_2O$ obtained by evaporating the yellow solutions at 370 K gives no precipitate with $BaCl_2$ and is evidently complex. Yellow $Ir_2(SO_4)_3aq$, made by treating Ir_2O_3 with H_2SO_4 in the absence of air, is also of unknown structure. Both Rh^{III} and Ir^{III} forms alums, however, with the sulphates of K, Rb, Cs, NH_4 and Tl^{I}

A hydrated palladium(II) nitrate, $Pd(NO_3)_2 \cdot 2\ H_2O$, obtained from a solution of the metal in nitric acid, has an infrared spectrum which indicates the existence of unidentate nitrato groups. A brown, volatile, anhydrous compound $Pd(NO_3)_2$, made by treating the dihydrate with liquid N_2O_5, is shown by the same method to contain bridging nitrato groups.

Clearly the platinum-group metals form very few compounds which can be considered as true salts.

ORGANOMETALLIC COMPOUNDS AND π-COMPLEXES

Platinum, unlike the other Group VIII Metals, forms some thermally stable alkyl compounds in which there is little or no π-bonding. Orange crystals of Me_3PtI are obtained from the reaction between $PtCl_4$ and MeMgI by treating the product with water and afterwards extracting with benzene. The solid is insoluble in water and is not attacked by concentrated acids and alkalis, but it reacts with potassium to give Pt_2Me_6 :

$$2Me_3PtI + 2\,K \rightarrow Me_3Pt{-}PtMe_3 + 2\,KI$$

Tetramethylplatinum has an interesting structure. The molecule is a cubic tetramer in which every platinum is octahedrally co-ordinated to methyl groups ; three of these are terminal groups and three bridging groups to other platinum atoms (Fig. 7.11). Octahedral arrangements in Pt^{IV} compounds are frequently preserved by such unusual types of co-ordination.

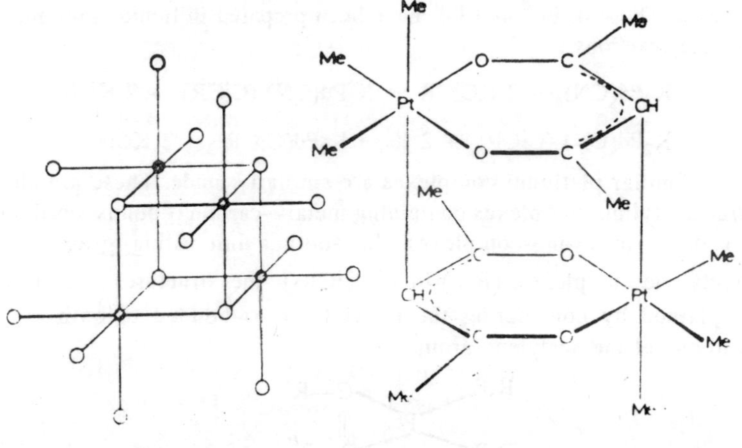

Fig. 7.11. Tetramethylplatinum : cubic structure of the tetrameric molecules.

Fig. 7.12. Dimeric molecule of trimethylplatinum acetylacetonate.

Trimethylplatinum acetylacetonate, made by the action of Me_3PtI on acetylacetone, has a dimeric molecule (Fig. 7.12). This dimer reacts with dipyridyl to give the monomeric Me_3Pt(dipy) $(O_2C_5H_7)$ in which the octahedral arrangement around the platinum is preserved through co-ordination, not, however, to the two oxygens of the β–diketone, but to the methylenic carbon atom (Fig. 7.13)

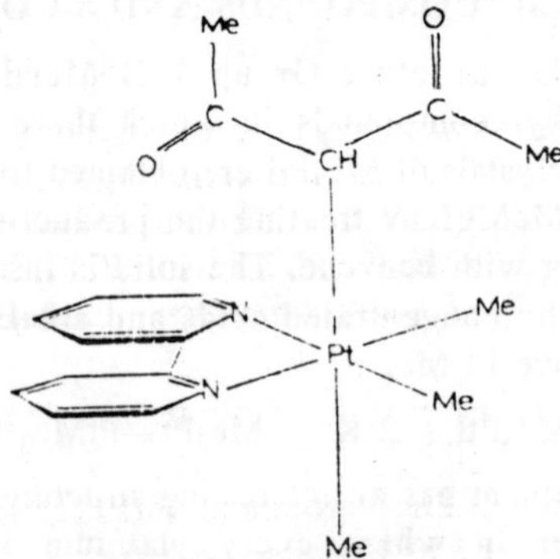

Fig. 7.13. Co-ordination of platinum to methylenic carbon atom in Me_3Pt (dipy)$(O_2C_5H_7)$.

Of the many olefin complexes formed by the platinum metals some have already been discussed. Complex acetylides are also known. Those of Pd^{II} and Pd^{0} have been prepared in liquid ammonia by the reactions :

$$K_2Pd(CN)_4 + 2\ KC{\vdots}CR \rightarrow K_2Pd(CN)_2(C{\vdots}CR)_2 + 2\ KCN$$

$$K_2Pd(CN)_2(C{\vdots}CR)_2 + 2\ K \rightarrow K_2Pd(C{\vdots}CR)_2 + 2\ KCN$$

Similar platinum complexes are similarly made. These are all true acetylido-complexes containing metal—carbon σ bonds similar to those in cyano-complexes. In another interesting group of acetylene complexes, $(R_3P)_2Pt\ (R_1C{\vdots}CR_2)$, the structures are best explained by considering the metal to be σ–bonded to both the carbons of the acetylene group :

$$\begin{matrix} R_3P & & & & C—R' \\ & \searrow & Pt & \swarrow & \| \\ R_3P & \nearrow & & \nwarrow & C—R'' \end{matrix}$$

The cyclopentadienyls of ruthenium and osmium resemble ferrocene in being stable in air and capable of aromatic substitution without decomposition. Yellow $(\pi\text{-}C_5H_5)_2Rh$ and colourless

$(\pi\text{–}C_5H_5)_2Os$ are made by treating $RuCl_3$ and $OsCl_4$ respectively with C_5H_5Na.

The ions $(C_5H_5)_2Rh^+$ and $(C_5H_5)_2Ir^+$, made by treating the acetylacetonates with C_5H_5MgBr, can be reduced to the biscyclopentadienyls by $NaBH_4$.

COMPLEXES

Zero oxidation states

This state is exhibited in the many carbonyls and also in complexes containing phosphine and arsine ligands. Examples are $Ru(CO)_3(PPh_3)_2$ and $Os(CO)_3(PPh_3)_2$. The Pd^0 compound $K_4Pd(CN)_4$ is obtained by reducing $K_2Pd(CN)_4$ with potassium in liquid ammonia. The brown diamagnetic $Pd(RNC)_2$ compounds (R = Ph, *p*-CH_3 . Ph and *p*-CH_3OPh) react with phosphines to give compounds such as $Pd(PPh_3)_3$ and $Pd(PPh_3)_4$. Similar complexes of Pt^0 are obtained by reducing $PtX_2(PR_3)_2$ with hydrazine in the presence of an excess of the phosphine. The compound $Pt(PF_3)_4$ is obtained by treating $PtCl_2$ with PF_3 at 370 K under pressure.

Unipositive states

The polymer $[RuCOBR]_n$ is obtained as one of the products of the reaction between $RuBr_3$ and CO under pressure. Bright-yellow $[Os(NH_3)_6]$ Br, with a magnetic moment of 1.5 μ_B, can be made by reducing $Os(NH_3)_6Br_3$ with potassium in liquid ammonia.

Rhodium(I) complexes are quite numerous ; most of the contain π–bonding ligands. The compound

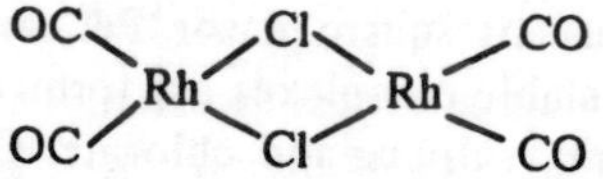

```
OC       Cl       CO
   \    /  \    /
    Rh        Rh
   /    \  /    \
OC       Cl       CO
```

reacts with many donor ligands to give monomeric $Rh(CO)_2ClL$ compounds. The compound $Rh(PPh_3)_3$ Cl can be made by treating $RhCl_3 . 3\ H_2O$ with PPh_3 in ethanol. It dissociates to give $Rh(PPh_3)_2Cl$, which catalyses the homogeneous hydrogenation of olefins and acetylenes ; the compound takes up H_2 to give *cis*-$H_2RhCl(PPh_3)_2$ to which the olefin attaches itself in the sixth co-ordination position and is hereby activated for hydrogenation to occur.

Iridium(I) is also a common species, but again nearly all the complexes contain π–bonding ligands. Like their Rh^1 analogues, many of the square d^8 complexes undergo oxidative addition reactions to give octahedral d^6 complexes, thus $Ir(CO)Cl(PPh_3)_2$ reacts with H_2, HCl and Cl_2 to give octahedral Ir^{III} complexes. The yellow compound of iridium made by reducing $[Ir(NH_3)_6]\ Br_3$ with potassium in liquid ammonia was first thought to be $Ir(NH_3)_5$ but it is diamagnetic and is more probably the Ir^{I} $HIr(NH_3)_5$.

BIPOSITIVE STATES

Ru^{II} and Os^{II}

The d^6 complexes of these are diamagnetic, octahedral, and usually inert. They are obtained by the direct action of the ligands on compounds in which the metal are in higher oxidation states :

$$RuCl_3 \xrightarrow[\text{evaporation on steam bath}]{\text{KCN solution}} K_4\ Ru(CH)_6$$

$$(NH_4)_2OsBr_6 \xrightarrow[540\ K]{\text{dipyridyl}} Os(dipy)_3BR_2$$

$$RuCl_3 \xrightarrow[\text{in ethanol}]{C_6H_4(AsMe_2)_2} Ru(diars)_2Cl_2$$

In these reactions the KCN, dipyridyl and diarsine all act as reducing agents.

Rh^{II} and Ir^{II}

There are few genuine examples of these states. Many of the compounds which have been reported have been shown later to be hydrido complexes of the +3 states.

Pd^{II} and Pt^{II}

These complexes are formally of d^8 configuration ; they are diamagnetic, and usually square planar. Pd^{11} and Pt^{11} are typical soft acids. Their most stable complexes are formed with donors such as phosphorus, arsenic, sulphur and chlorine. Oxygen-donor ligands produce only a few unstable complexes, although there are many complexes with nitrogen ligands.

There are a few octahedral complexes of Pt^{II} such as $[PtCl_5NO]^{2-}$ and $[PtCl(NO)(NH_3)_4]^{2+}$.

TERPOSITIVE STATES

Ru^{III} and Os^{III}

The d^7 complexes are all low-spin, with one unpaired electron. Ruthenium(III) complexes are much more common than those of osmium(III).

Rutenium halides react with ammonia to give hexa-amines, which are changed to acidopenta-ammines by boiling with acid :

$$RuCl_3 + 6\ NH_3 \rightarrow [Ru(NH_3)_6]Cl_3 \xrightarrow{HCl\ (boil)} [RuCl(NH_3)_5]Cl_2$$

Complexes of osmium(III) with nitrogen ligands are not common, but $[Os(NH_3)_6]^{3+}$, $[OsBr(NH_3)_5]^{2+}$ and $[Os(dipy)_3]^{3+}$ occur.

Only a few Ru^{III} and Os^{III} complexes with oxygen donors are known—mainly β-diketone and oxalato chelates.

In keeping with the soft acid character of these states they form stable complexes with phosphines and arsines. Examples are $RuX_3(PPh_3)_3$ and $[Os(diars)_2X_2]^+$.

Rh^{III} and Ir^{III}

There are many stable complexes of these d^6 states.

The cationic complexes resemble those of Co^{III} (also d^6), and the structures, where known, are octahedral. They are all diamagnetic ; even a weak-field ligand like F^- causes spin-pairing in the $RuF_6^{\ 3-}$ ion.

There is a well-defined $Rh(H_2)_6)^{3+}$ ion in rhodium alums and in the perchlorate $Rh(H_2O)_6(ClO_4)_3$; this is notable because aquo ions of this type are uncommon among second and third row d-block elements. Salts of *cis*-$[Rhen_2Cl_2]^+$ have been resolved into optical isomers. The $[Rh(dipy)_3]X_3$ salts (X = Cl, br, I, SCN, ClO_4) are yellow because of charge-transfer bands in the blue region of the spectrum.

Rhodium(III) forms many more anionic complexes than does cobalt(III) ; they are usually more labile than the cationic and neutral complexes.

Palladium and platinum

The +3 state probably does not exist in the true sense in either of these metals, for instance, PdF_3 is actually $Pd^{II}[Pd^{IV}F_6]$ and $PtBr_3en$ contains equal numbers of $PtBR_2en$ and $PtBr_4en$ groupings.

Quadripositive states

Ru^{IV}, Rh^{IV}, Pd^{IV}, Os^{IV} and Ir^{IV} occur principally in complex halides but there is a greater range of Pt^{IV} complexes, some of which can be made by *trans* addition :

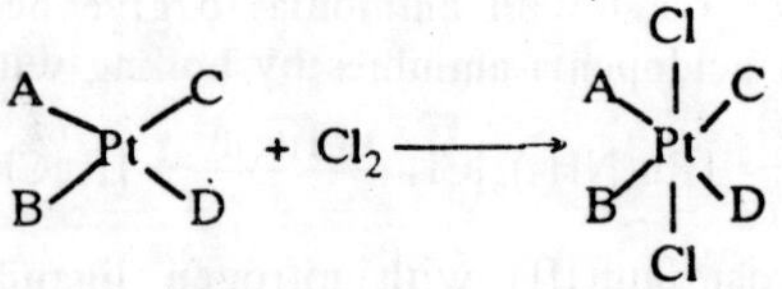

In the product the equatorial positions of the octahedral Pt^{IV} complex have the same arrangement as in the square Pt^{II} complex. The Pt^{IV} complexes are invariably octahedral ; they provide a particularly extensive range of ammines, from $Pt(NH_3)_6^{4+}$ through $PtX(NH_3)_5^{3+}$, all the way to PtX_6^{2-} (X = Cl, Br, SCN, NO_2. Ethylenediamine, hydrazine and hydroxylamine are other nitrogen donors with appear in these complexes.

HYDRIDOCOMPLEXES

When ligands which exert a strong field are attached to a transition metal it acquires some of the σ–bonding character of a B sub-group metalloid. Thus it can form metal—metal bonds, as in $Mn_2(CO)_{10}$, and strong σ–bonds to hydrogen, as in carbonyl hydrides, e.g. $MnH(CO)_5$, cyclopentadienyl hydrides, e.g. $ReH(C_5H_5)_2$, and carbonylcyclopentadienyl hydrides, e.g. $MoH(C_5H_5)(CO)_3$.

The platinum metals form a particularly interesting series of molecular hydrides in which stabilisation is due to tertiary phosphines and amines. Examples are :

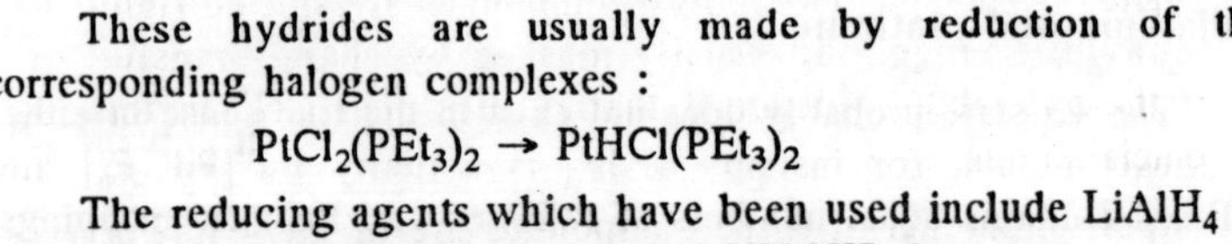

These hydrides are usually made by reduction of the corresponding halogen complexes :

$$PtCl_2(PEt_3)_2 \rightarrow PtHCl(PEt_3)_2$$

The reducing agents which have been used include $LiAlH_4$ in tetrahydrofuran, H_3PO_2 in alcohol and NH_2NH_2 in water.

In these compounds hydrogen acts as an anionic ligand, the metal retains the same charge number, co-ordination number and stereochemistry as it has in the halogen compound from which the hydride is made. Thus *trans*-$PtHBr(PEt_3)_2$ has the slightly distorted square planar structure (Fig. 7.14).

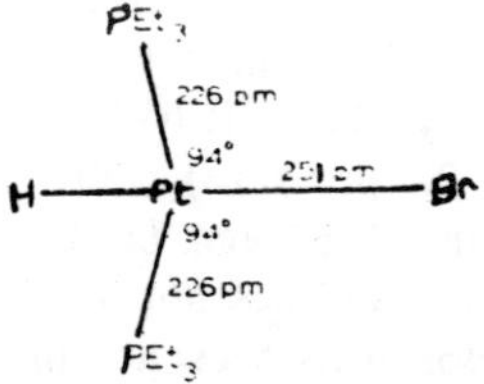

Fig. 7.14. Bond lengths and bond angles in *trans*-$PtHBr(PEt_3)_2$.

The chemical shifts in the proton magnetic resonance spectrum are very large (20–30 p.p.m.) – well removed from those due to organic substituents on the phosphorous atoms. IN a compound such as $PtHBr(PEt_3)_2$ the hydrogen resonance is split into a triplet by the two equivalent ^{31}P nuclei and there is further large splitting by ^{195}Pt.

The infrared absorption spectra show a strong, sharp band due to metal—hydrogen bond stretching, but the position of the band varies a great deal from one compound to another (1726–2242 cm^{-1}). In platinum complexes of the *trnas*-$PtHX(PEt_3)_2$ type the M—H stretching frequency is reduced by X ligands of increasing *trans* effect.

X	NO_3	Cl	BR	I	NO2	SCN	CN
M—H stretching frequency/cm^{-1}	2241	2183	2178	2156	2150	2112	2041

The hydrogen itself exerts a strong *trans* effect. In the process :

$$trans\text{-}PtXCl(PEt_3)_2 + py \rightarrow PtXpy(PEt_3)_2^+ + Cl^-$$

the rate of reaction is 10^6 times faster for X = H than for X = Cl. The hydrogen produces a strong ligand field at the metal ; the d—d transition bands in the spectra of these compounds are shifted right into the ultraviolet range, and usually masked by charge-transfer bonds, but the available evidence is that H^- lies near CN^- in the spectrochemcial series.

The dipole moments of these compounds are of interest. For *trans*-$PtHCl(PEt_3)_2$ the moment is 4.2 D. As the usual moment associated with a transition-metal to chlorine bond is about 2 D the M—H moment appears to be about 2.5 D, the with H positive, in conflict with the idea that the hydrogen is co-ordinated as hydride ion. However, some of the moment is due to the distortion from square symmetry (Fig. 7.14) which places positive P atoms towards the hydrogen ; furthermore the presence of the hydrogen has the effect of lengthening the metal—chlorine bond.

Thermal stabilities of analogous hydridocomplexes in any group of transition elements usually rise with the atomic mass of the element. Thus in the nickel group $PtHCl(PEt_3)_2$ is stable enough to be distilled at 400 K (1.5 Pa pressure) ; the palladium compound $PdHCl(PEt_3)_2$ is rather unstable even in the solid state ; but the corresponding nickel compound has not been isolated, although its n.m.r. spectrum has disclosed its presence in the solution made by reducing $NiCl_2(PEt_3)_2$ with $LiAlH_4$. The order of thermal stability of these transition-metal hydridocomplexes is thus the reverse of that found for hydrides of a B-sub-group :

Ni complex < Pd complex < Pt complex cf. NH3 > PH3>AsH3> SbH3

The arsine- and phosphine-stabilised hydrodocomplexes of the platinum metals are usually octahedral or square, but the trigonal bipyramidal $RhH(CO)(PPh_3)_3$ has been made by reducing the square planar $RhCl(CO)(PPh_3)_2$ with hydrazine in alcohol.

The π–bonding ligands PH_3 and AsR_3 which stabilise the metal—hydrogen σ bonds are also effective in stabilising metal—carbon σ bonds and there are some organometallic compounds similar to these hydrides. *Trans*-$PtHCl(PEt_3)_2$ reacts reversibly with ethylene to give *trans*-$PtClEt(PEt_3)_2$.

CHAPTER 8

Copper, Silver and Gold—Group IB

INTRODUCTION

Traditionally the coinage metals because of their resistance to corrosion, Cu, Ag and Au form the sub-group which marks the ends of the respective 'd-block' series. The atoms have the configuration $(n—1)d^{10}\,ns^{1}$ but the metals are nevertheless considered as transitional because ions with incomplete d shells can be formed by all three. The first ionisation energy (Table 8.1) is much greater than that of the preceding alkali metal with its $(n—1)\,p^{6}\,ns^{1}$ configuration, because the tend electrons are rather ineffective in shielding the outer s electron from the coulombic field of the nucleus. However, d electrons can either be lost in the formation of 2+ ions (Cu^{2+} and Ag^{2+}) since the second ionisation energies are not high, or can take part in covalent bonding to give formal oxidation states as high as + 3 (particularly Au^{III}). The compounds of the metals in these higher oxidation states are coloured and paramagnetic like those of typical transition metals.

Metallic binding in the elements is strong because electrons of the d shells are involved as well as the outer s electrons; consequently the *m.p.* and enthalpies of sublimation are high ; a further consequence is that conversion of the metals to aquated cations is energetically unfavourable, and the metals do not corrode. Conversely, they are easily deposited electrolytically from aqueous solutions of their cations.

Again because of the high effective nuclear charges acting on the outer electrons, the internuclear distances in the solids are low, and the metals have high densities, particularly Au, in which, following the lanthanide contraction, the metallic radius is almost identical with that of Ag. The ions are also small, for example the radius of Cu^+ is only 96 pm compared with 133 pm for K^+. Thus the compounds tend to have high lattices energies, and many of them are insoluble in water as a consequence.

TABLE 8.1 ATOMIC PROPERTIES OF THE ELEMENTS

	Cu	Ag	Au
Z	29	47	79
Electron configuration	[Ar] $3d^{10}\,4s^1$	[Kr] $4d^{10}\,5s^1$	[Xe] $4f^{14}\,5d^{10}\,6s^1$
$I(1)$/kJ mol^{-1}	745	731	889
$I(1)$/kJ mol^{-1}	1958	2072	1980
Ionic radii M^+/pm	128	143	144
Metallic radii/pm	96	126	137

TABLE 8.2 PHYSICAL PROPERTIES OF THE ELEMENTS

	Cu	Ag	Au
ρ/g cm^{-3}	8.93	10.5	19.3
M.p./K	1357	1234	1338
B.p./K	2868	2485	3000

THE ELEMENTS

The metals all occur native, gold almost exclusively so. Copper makes up about 7×10^{-3}% of the lithosphere mainly as copper pyrites $CuFeS_2$, cuprite, Cu_2O, and malachite, $Cu_2(OH)_2CO_3$; silver about 2×10^{-5}% mainly as argentite, Ag_2S, horn silver, AgCl, and pyrargyrite, Ag_3SbS_3; gold about 5×10^{-7}% mainly as metal. Their extraction from ores is chemically relatively easy.

Copper

Oxide and carbonates ores are readily reduced by heating wi... coke and a flux :

$$CuCO_3 \xrightarrow{heat} CuO \xrightarrow{C} Cu$$

The main source is copper pyrites which is relieved of volatile arsenic and antimony by roasting, after which it is slagged and reduced.

(i) Partial oxidation : $CuFeS_2 \xrightarrow{O_2} Cu_2S + FeS + SO_2$

(ii) Slagging off iron :

$$Cu_2S + FeS + SiO_2 \rightarrow \underset{\text{(less dense)}}{FeSiO_3 \text{ slag}} + \underset{\text{(more dense)}}{Cu_2S \text{ liquid sulphide}}$$

(iii) Oxidation—reduction :

$$Cu_2S \xrightarrow{O_2} Cu_2O + Cu_2S \xrightarrow{\text{heat alone}} Cu + SO_2$$

Copper is also extracted from its ores by leaching with a solution of H_2SO_4 and $Fe_2(SO_4)_3$, which converts the CuS or $CuCO_3$ to $CuSO_4$ solution. Copper is refined electrolytically, the silver and gold present separating as an anode sludge.

Silver and gold

Whether present as metal or compound, these elements can be extracted from the finely divided ore by means of aqueous sodium cyanide. The cyanide ion reduces the oxidation potential of the noble metal so that atmospheric oxygen brings it into solution as a soluble complex :

$$4\,Ag + 8\,NaCN + H_2O + O_2 \rightarrow 4\,NaAg(CN)_2 + 4\,NaOH$$

$$Ag_2S + 4\,NaCN \rightarrow 2\,NaAg(CN)_2 + Na_2S$$

The sodium sulphide is largely oxidised to Na_2SO_4 by air and the reverse reaction is thereby impeded. The silver and gold are precipitated from their solution as cyano-ions, after this has been filtered, by the addition of zinc :

$$2\,Ag(CN)_2^- + Zn \rightarrow 2\,Ag + Zn(CN)_4^{2-}$$

$$2\,Au(CN)_4^- + 3\,Zn + 4\,CN^- \rightarrow 2\,Ag + 3\,Zn(CN)_4^{2-}$$

Silver and gold are recovered during the purification of lead and nickel.

Uses

World production of copper is about 6 million tons per annum. It is therefore a metal of great commercial importance, being used particularly as an electrical conductor, in casting alloys and in coinage. The biochemical behavior of copper is of interest. It is a constituent of many oxidases—enzymes which catalyse redox reactions—in which it functions through the shifting of the Cu^{II}/Cu^{I} equilibrium during ligand replacements. Copper deficiency in plants gives rise to die-back, inability to form seeds, chlorosis, and the reduction of photosynthetic activity. Copper deficiency in soils is made good by spraying with aqueous $CuSO_4$.

World production of silver is about 10 thousand tons per annum. Its alloys with copper, particularly Sterling silver (92.5% Ag, 7.5% Cu), are used in tableware. The metal is also used in silvering mirrors and, increasingly, in storage batteries.

The annual world production of gold is about 15 hundred tons—three quarters of this in South Africa.

About half of the world's gold is used for adjusting trade balances. Its use in jewellery is well known, and it is becoming of increasing importance in electronics.

PROPERTIES

The metals all have the c.c.p. structure and are rather soft when pure, presumably because of the regularity of the packing in the lattice. They are particularly malleable and ductile and are excellent conductors of electricity and heat. The boiling points are high, in excess of 2400 K. The elements form alloys with one another and with many other metals; of particular importance are brass (copper—zinc) and bronze (copper—tin), which are much harder than pure copper.

REACTIONS

Gold is rather unreactive, though it is attacked by BrF_3 and aqua regia in the cold and also by warm Cl_2 and Br_2. Copper and silver are more reactive; the former is converted to CuO by oxygen at red heat. The halogens react with the heated metal to give dihalides, and sulphur combines with it to give a product of approximate composition Cu_2S. Non-oxidising acids have no action, but HNO_3 and H_2SO_4 dissolve copper, being themselves reduced. Silver is less reactive than copper except towards S and H_2S which both convert it to black Ag_2S. Alkaline cyanide solutions containing oxidising agents such as H_2O_2 are good solvents for Ag and Au, which react to form the $Ag(CN)_2^-$ and $Au(CN)_4^-$ ions respectively.

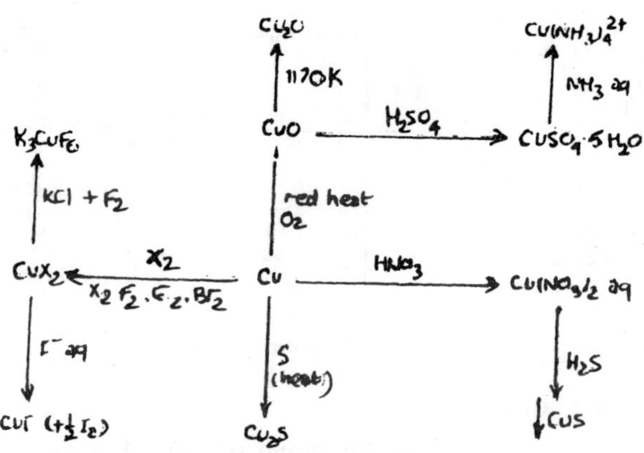

Fig. 8.1. Reactions of copper.

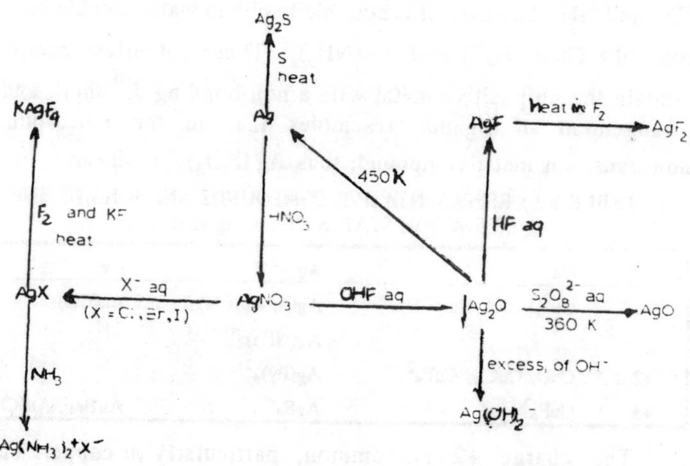

Fig. 8.2. Reactions of silver.

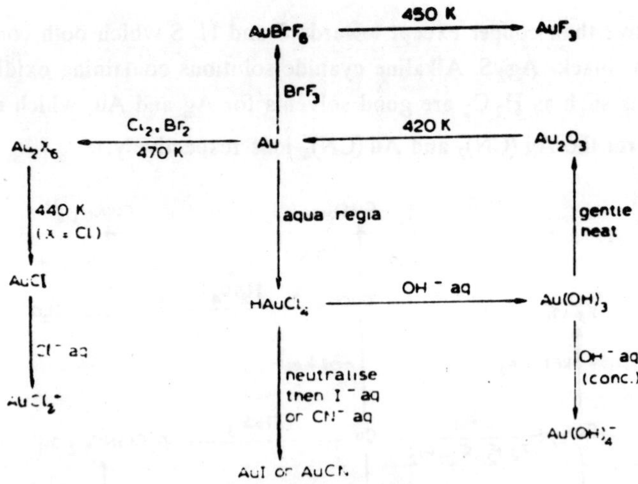

Fig. 8.3. Reactions of gold.

Oxidation States

Charge numbers in the sub-group are limited to +1, +2 and +3. There is no evidence for the zero state or negative states. In the simple compounds of the unipositive metals the bonding is rather covalent in character; lattice energies are therefore enhanced and solubilities are low. In the presence of complexing agents such as CN^- and NH_3, however, it is possible to obtain water-soluble complex ions like $Cu(CN)_4^{3-}$) and $Ag(NH_3)_2^+$. These colourless complexes contain the unipositive metal with a non-bonding d^{10} shell, and the arrangement of ligands resembles that in the corresponding non-transition metal compound; thus $Ag(NH_3)_2^+$ is linear.

TABLE 8.3 : REPRESENTATIVE COMPOUNDS AND IONS OF THE OXIDATION STATES OF Cu, Ag AND Au

	Cu	Ag	Au
+1	Cu^2O, CuI, $Cu(CN)_4^{3-}$	AgCl, $Ag(CN)_2^-$	$Au(CN)_2^-$
		$Ag(NH_3)_2^+$	
+2	CuO, $CuCl_2$, CuF_4^{2-}	$Ag(Py)_4^{2+}$	
+3	CuF_6^{3-}	AgF_4^-	$AuBr_4^-$, $Au(CN)_4^-$

The charge +2 is common, particularly in copper. The d^9 arrangement in Cu^{II} gives rise to square arrangements of ligands around the metal atom or to distorted octahedral arrangements in which two ligands are attached, by somewhat longer bonds, above and below the plane of the other four. (Compounds of Ag^{II} are mostly strong oxidising agents; it has not yet been possible to make a compound of Au^{II}.

All three metals form M^{III} complexes. Pale-green K_3CuF_6 has a paramagnetic moment of 2.8 μ_B, consistent with the presence of a d^8 configuration containing two unpaired electrons. Gold (III) complexes are common; they are oxidising agents in which the Au is usually at the centre of a square formed by four ligands.

The free energies of the oxidation states of Cu, Ag and Au relative to the metals in aqueous solutions at pH 0 are represented graphically in Fig. 8.4. Although *I*(2) for Cu is rather high, the enthalpy of hydration of Cu^{2+} is so much more negative than that of Cu^+ that the dipositive ion is much the more stable in aqueous solution. Thus soluble Cu^+ salts disproportionate in water. Silver (I) salts do not do so however.

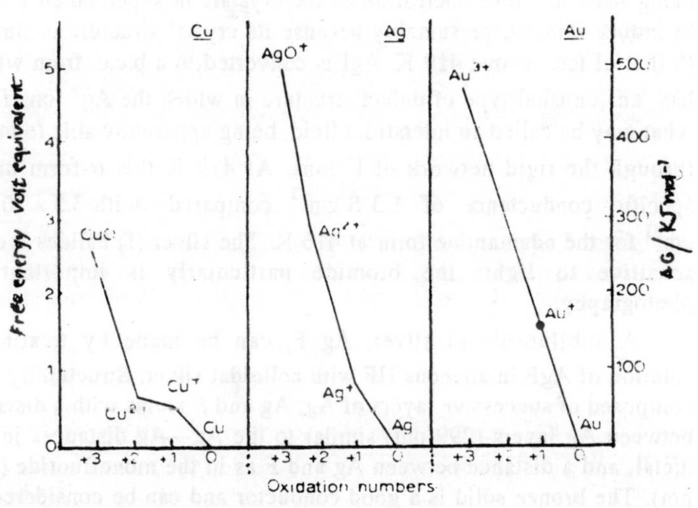

Fig. 8.4. Free energies of the oxidation states of the group IB elements relative to the metals at pH = 0.

Halides

White CuCl or CuBr can be precipitated from aqueous $CuCl_2$ or $CuBr_2$ by a variety of reducing agents including aqueous SO_2, $SnCl_2$ and $N_2H_5HSO_4$, but CuI is obtained merely by adding I^- to a solution of Cu^{2+} salt;

$$2\,Cu^{2+} + 2\,I_2 \rightarrow 2\,CuI + I_2$$

A reaction which is used for the volumetric determination of Cu^{2+}. All three copper (I) halides have the zinc blende structure, and they have considerable covalent character, the experimental values for the lattice energies being markedly larger than the values calculated on the basis of ionic models. The halides are moderately soluble in solutions containing the corresponding halide ion because complex halides are formed.

The silver (I) halides, white AgCl, cream AgBr and pale yellow AgI are made by precipitation reactions, but the more soluble AgF is best prepared by dissolving Ag_2O in HF and evaporating until yellow AgF crystallises. AgCl, AgBr and anhydrous AgF have the NaCl structure but AgI has both wurtzite and blende structures, the former being used to cause nucleation of ice crystals in supercooled clouds to induce rainfall, presumably because its crystal structure is similar to that of ice. Above 419 K, AgI is converted to a b.c.c. from which has an unusual type of defect structure in which the Ag^+ ions form what may be called an interstitial fluid, being apparently able to move through the rigid network of I^- ions. At 419 K this α-form has a specific conductance of 1.3 S cm^{-1} compared with 3.5×10^{-4} S cm^{-1} for the adamantine form at 416 K. The silver (I) halides are all sensitive to light; the bromide particularly is important in photography.

A subfluoride of silver, Ag_2F, can be made by treating a solution of AgF in aqueous HF with colloidal silver. Structurally it is composed of successive layers of Ag, Ag and F atoms with a distance between Ag layers (299 pm) similar to the Ag—Ag distances in the metal, and a distance between Ag and F as in the monofluoride (245 pm). The bronze solid is a good conductor and can be considered as intermediate between a metal and a salt.

Gold (I) chloride can be made by heating $AuCl_3$ to 430 K. It is decomposes at a slightly higher temperature.

Anhydrous CuF_2, made the elements, is an ionic compound with a distorted rutile structure. But the more covalent anhydrous $CuCl_2$ and $CuBr_2$ have polymeric chain structures formed by planar CuX_4 groups sharing opposite edges, the chains being packed so that each Cu has two more halogens at a rather greater distance. In $CuCl_2$ the Cu—Cl distances are 230 pm within the chain and 295 pm between chains. In $CuCl_2 \cdot 2H_2O$ however there are finite groups with the structure :

Cl OH$_2$
Cu
H$_2$O Cl

Silver (II) fluoride, made by the action of F_2 on Ag or AgCl; although it is thermally stable, the vapour pressure of the fused compound being only 10 kPa at 970 K, is a powerful fluorinating agent, particularly useful in the preparation of fluorocarbons from hydrocarbons.

AuF_3, made by treating Au with BrF_3 and then warming the $AuBrF_6$ so formed, has a polymeric structure in which each Au is surrounded by a square of F atoms and each square is linked to the next by *cis* bridges to give an infinite hexagonal helix. It is a vigorous fluorinating agent.

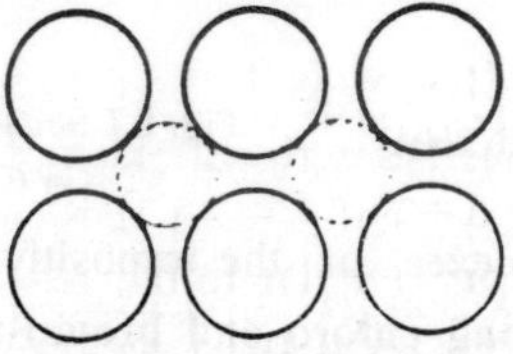

Fig. 8.5. Planar molecule of Au_2Cl_6.

Gold (III) chloride contains planar Au_2Cl_6 molecules in both the solid and the vapour (Fig. 8.5). It is best made by direct chlorination at 470 K.

Complex Halides

The halogen complexes of copper (I), with general formulae $M_2^1CuX_3$ and $M^1Cu_2X_3$, contain 4-co-ordinate Cu^I, the former containing, single chains, and the latter double chains of CuX_4 tetrahedra sharing corners. In Cs_2AgI_3 there are chains of AgI_4 sharing corners, but in $Me_4NAg_2I_3$ similar tetrahedra are joined by sharing edges.

Among the halide complexes of the dipositive metals, $K_2CuCl_4 \cdot 2H_2O$ (Fig. 8.6) has a tetragonal unit cell with features deriving from both $CuCl_2 \cdot 2H_2O$ and $CuCl_2$. In anhydrous Cs_2CuCl_4 the $CuCl_4^{2-}$ ion is not planar but a flattened tetrahedron; in aqueous solution the absorption spectrum is different from that of the solid, possible because the anion reverts to a planar configuration on hydration.

The Cu^{II} complex $CsCuCl_3$ consists of Cs^+ ions and infinite chain ions $(CuCl_3^{2-})_n$ (Fig. 8.7).

The compound formulated $CsAuCl_3$, however, contains both $[AuCl_2]^-$ and $[AuCl_4]^-$ ions, it contains both Au^I and Au^{III}. There is an isostructural compound $Cs_2[AgCl_2][AuCl_4]$.

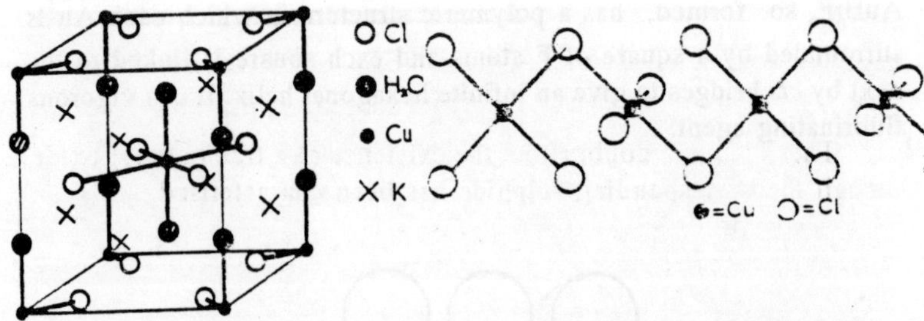

Fig. 8.6 Structure of $K_2CuCl_4 2H_2O$

Fig. 8.7 Spiral arrangement of atoms in $(CuCl_3^{2-})n$ ions

Of halogen complexes of the terpositive metals, K_3CuF_6 is known, but corresponding chloro and bromo-complexes have not been made. As so often happens in trasition-metal chemistry the F^- ion is the only halide ion capable of stabilising the highest oxidation state. The paramagnetic moment of K_3CuF_6 indicates the presence of two unpaired electrons, suggesting that copper (III) has the $t_{2g}^6 e_g^2$ configuration.

Gold (III) halogen complexes are particularly stable. The $AuBr_4^-$ ion in $KAuBr_4 \cdot 2H_2O$ is square, as is the AuF_4^- ion present in $KAuF_4$.

OXIDES

Copper (I) oxide occurs as the red mineral cuprite. It is best made in the laboratory by reducing Fehling's solution—an alkaline solution of a Cu^{II} salt containing just sufficient tartrate to keep the copper in solution—with an aldehyde. A hydrated Ag_2O is obtained as a dark-brown precipitate when aqueous OH^- is added to aqueous Ag^+. It is difficult to remove all the water without causing decomposition but anhydrous Ag_2O can be obtained by heating Ag powder in oxygen. The oxides Cu_2O and Ag_2O are isomorphous and of unusual structure. The metal atoms have two collinear bonds and the oxygen four tetrahedral bonds in a cubic structure similar to that of β-cristobalite. The low co-ordination 4 : 2 indicates covalent character. The structure represented in Fig. 8.8 is incomplete because an identical framework, in which the oxygens marked B take up positions A, interpenetrates it—a structure unique in crystal chemistry.

Ag_2O is decomposed completely into silver and oxygen at 570 K under normal pressure. It gives an alkaline reaction with water; the solubility, though slight, is in conformity with the low lattice energy. Cu_2O has much greater thermal stability, a much more negative enthalpy of formation (ΔH_f is—170 kJ mol^{-1} compared with—30 kJ mol^{-1} for Ag_2O), and a lower solubility than Ag_2O.

There is some doubt about the existence of a true gold (I) oxide though the corresponding sulphide has been characterised.

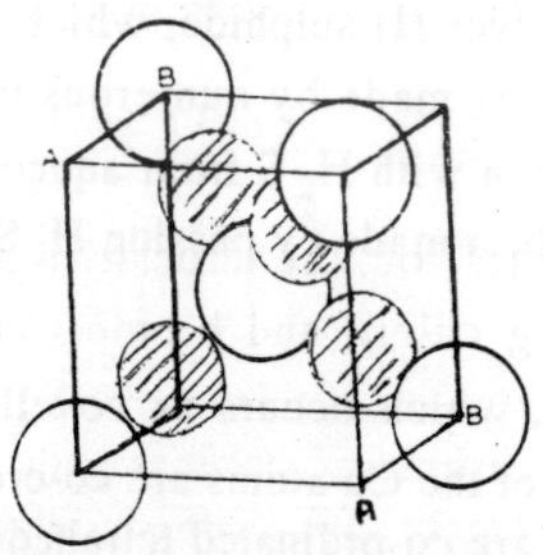

Fig. 8.8. Cu_2O structure. An identical framework in which positions A move back to occupy positions B interpenetrates the framework which is illustrated.

Copper (II) oxide can be made by heating powdered copper in air or oxygen or by decomposition of the pale-blue Cu $(OH)_2$ made by adding OH^- aq to a solution of Cu^{2+} in just sufficient aqueous NH_3 to hold the cations in solution. The structure of CuO resembles that of PdO; it involves 4 : 4 coordination, with coplanar bonds around the copper atoms and a tetrahedral arrangement of copper atoms around the oxygens. The Cu—O distance of 195 pm indicates a large degree of covalency.

A compound with the composition AgO can be obtained as a black precipitate by treating aqueous $AgNO_3$ with alkaline $S_2O_8^{2-}$. It is not paramagnetic as a true Ag^{II} compound should be. Neutron diffraction studies have shown that it contains Ag^{I} co-ordinated to two oxygens and Ag^{III} co-ordinated to four oxygens. The compound is a semiconductor and a very strong oxidising agent. It decomposes at 370 K to give Ag_2O and oxygen, and dissolves in aqueous HNO_3, with some loss of oxygen, to give a paramagnetic solution which presumably contains Ag^{2+}.

Addition of OH^- to aqueous $AuCl_4^-$ gives a light-brown precipitate of composition $Au(OH)_3$ which is converted to brown Au_2O_3 by heating to constant weight at 420 K. The anhydrous oxide is soluble in concentrated mineral acids and also in hot aqueous alkalis; yellow needles of $KAuO_2 \cdot 3H_2O$ can be crystallised from a solution of Au_2O_3 in aqueous KOH.

SULPHIDES

Black Cu_2S occurs as chalcocite. It can be made by passing dry H_2S over Cu at 700 K. Silver (I) sulphide, which is also black and also occurs naturally, can be made by numerous methods but most conveniently by precipitation with H_2S from aqueous Ag^+ solutions. Black gold (I) sulphide is best made by passing H_2S into an acidified solution of $Au(CN)_2^-$.

The compound CuS, which occurs as covellite, is not a true Cu^{II} compound. One-third of the Cu atoms are co-ordinated trigonally to S atoms and two-thirds are co-ordinated tetrahedrally, furthermore two-thirds of the sulphur is present as S_2 groups. When covellite is heated with sulphur a darkpurple compound is obtained with the approximate composition CuS_2 and the pyrites structure. Gold (III) sulphide, made by passing H_2S into a solution of Au_2Cl_6 in ether, is a black, insoluble solid.

OTHER BINARY COMPOUNDS

A nitride Cu_3N can be made by heating CuF_2 with NH_3. It has a reversed ReO_3 structure with Cu—N distances of 190 pm. It is evidently a true Cu^I compound; it reacts with acids to give a Cu^{2+} salt, an NH_4^+ salt and a precipitate of copper :

$$2\,Cu_3N + 8\,H^- \rightarrow 2\,NH_4^+ + 3\,Cu^{2+} + 3\,Cu.$$

There is also an explosive azide CuN_3 made by reducing aqueous Cu^{2+} with HSO_3^- in the presence of N_3^- ions. An explosive silver (I) azide is precipitated by hydrazine from aqueous $AgNO_3$. There is some doubt about the existence of a true nitride or azide of gold ; the grey, explosive compound made by the action of NH_3 on aqueous $HAuCl_4$ contains $Au_2O_3 \cdot 3\,NH_3$ and $HNAuCl \cdot NH_3$.

The explosive solid Cu_2C_2 is obtained as a dark-red precipitate when C_2H_2 is passed into a solution of a Cu^+ salt in aqueous ammonia. Yellow Ag_2C_2 is made similarly from acetylene and ammoniacal $AgNO_3$. The dry solid detonates violently at 400 K.

OXOACID SALTS

Copper (I) sulphate is the only ionic Cu^I compound. It is made by heating Cu_2O with dimethyl sulphate :

$$Cu_2O + Me_2SO_4 \rightarrow Cu_2SO_4 + Me_2O$$

It is decomposed immediately in water :

$$Cu_2SO_4 \rightarrow Cu + CuSO_4$$

This is to be expected from the redox potentials of the Cu^+/Cu and Cu^{2+}/Cu couples. The Cu^+ ion is, however, stabilised by complexing and appears as colourless crystals of $|Cu(NH_3)_2|SO_4$; these are produced when ethyl alcohol is added to a solution made by dissolving Cu_2O in an aqueous solution of ammonium sulphate and ammonia.

Copper (II) sulphate pentahydrate, $Cu(H_2O)_4SO_4 \cdot H_2O$, has four water molecules and an oxygen from each of two SO_4^{2-} anions octahedrally arranged about the Cu^{2+} cation. The fifth water molecule is hydrogen-bonded to oxygen atoms. The positions of all the water molecules and hydrogen bonds have been determined by neutron refraction. The H—O—H angles of all the water molecules are near the tetrahedral angle, the O—H....O angles between 154—176°, and the O—O—O angle between 105—130°; so that some of the hydrogen bonds are bent, one by as much as 26°. The corresponding ammine hydrate, $Cu(NH_3)_4SO_4 \cdot H_2O$, is known, but not a pentaammine.

Copper (II) nitrate has the hydrates $Cu(NO_3)_2 \cdot 6H_2O$ and. $Cu(NO_3)_2, 9H_2O$Attempts to dehydrate them produce basic salts. However, copper reacts vigorously with N_2O_4 in ethyl acetate, and a solid $Cu(NO_3)_2 \cdot N_2O_4$ can be crystallised from the solution. Gentle heating distils off the N_2O_4 leaving the green, volatile $Cu(NO_3)_2$. This compound is monomeric in the vapour, the Cu atom being co-ordinates to NO_3 groups acting as bidentate ligands.

The most important soluble salt of silver, $AgNO_3$, can be prepared by dissolving Ag in HNO_3. It is stable up to 620 K but at 700 K it decomposes into the metal, nitrogen, nitrogen oxides and oxygen. Unlike the halides, the pure nitrate is not photosensitive.

Copper (II) acetate monohydrate is interesting for its magnetic properties.

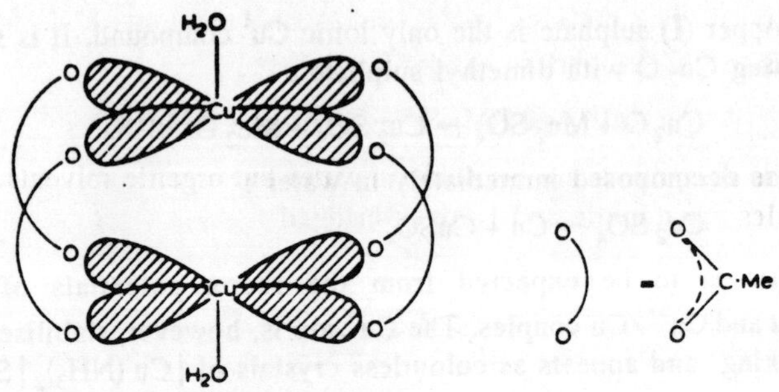

Fig. 8.9. Dimer of copper (II) acetate monohydrate, illustrating δ-bonding.

The moment decreases as the temperature falls, suggesting a temperature dependent equilibrium between the diamagnetic d^{10} and the paramagnetic d^9 configurations, with a Cu—Cu bond energy of only about 4 kJ mol^{-1}. The compound has a dimeric structure (Fig. 8.9). The copper atoms have square pyramidal arrangements of oxygens around them. Interaction between the two copper atoms is pictured as being due to the face-to-face overlap of the $d_{x^2-y^2}$ orbitals, shown in the diagram. This type of bonding, called δ-bonding, though it is extremely weak compared with δ-and π-bonding, can lead to diamagnetism through the coupling of spins.

Pale-yellow Ag_2CO_3 can be precipitated from aqueous $AgNO_3$ with alkali carbonates. In contrast only basic carbonates can be precipitated from Cu^{2+} solutions, an indication that Ag_2O is more basic than CuO.

Silver perchlorate is not only deliquescent and very soluble but forms a monohydrate. It also dissolves in benzene and toluene, recrystallising from the latter as $AgClO_4 \cdot C_7H_8$. Solubility in organic solvents is thus shown to be an unreliable guide to covalent character, for silver perchlorate is considerably ionised both in water and nitromethane.

Oxoacid salts of gold are uncommon; one of the more stable, yellow $Au_2(SeO_4)_3$, crystallises from a solution of gold in hot selenic acid.

ORGANOMETALLIC COMPOUNDS

Gold resembles platinum in forming numerous δ-bonded alkyl derivatives. The halides $AuCl_3$ and $AuBr_3$ react with Grignard reagents to give colourless solids, R_2AuX :

$$AuBr_3 + 2\,RMgBr \rightarrow R_2AuBr + 2\,MgBr_2$$

The compounds are insoluble in water but organic solvents. The molecules are dimeric and halogen-bridged :

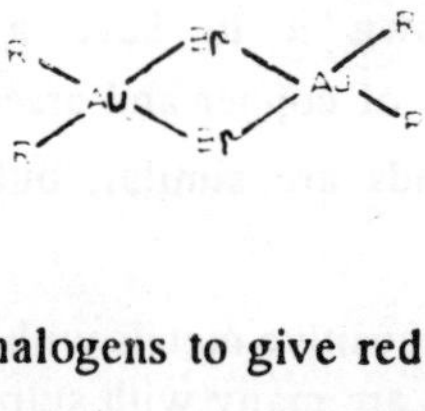

They react with halogens to give red monoalkyl derivatives :

$$R_2AuBr + Br_2 \rightarrow RAuBr_2 + R\,Br$$

The cyanides R_2AuCN are tetrameric :

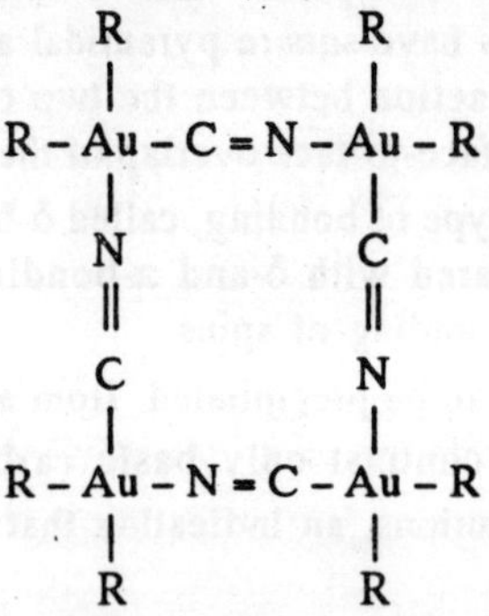

```
  R          R
  |          |
R – Au – C ≡ N – Au – R
  |          |
  N          C
  ||         ||
  C          N
  |          |
R – Au – N ≡ C – Au – R
  |          |
  R          R
```

Aryl compounds cannot be made from Grignard reagents, but yellow solids of the general formula $ArAuCl_2$ are prepared by dissolving $AuCl_3$ in the aromatic hydrocarbon :

$$AuCl_3 + C_6H_6 \rightarrow C_6H_5AuCl_2 + HCl$$

The known alkyl and aryl compounds of silver are of low thermal stability. The usual method of preparation is the treatment of $AgNO_3$ with a tetraalkyl or aryl of lead in ethanol :

$$Ag^+ + R_4Pb \rightarrow R_3Pb^+ + RAg$$

COMPLEXES

The +1 state

The three metals of the sub-group form ammines containing linear $M(NH_3)_2^+$ ions. There are similar alkylamine and pyridine complexes. Phosphine, and substituted phosphines, and arsines react with monohalides of the metals to form rather unstable complexes :

H_3PCuBr $\quad$ Et_3PCuI $\quad$ Et_3AsCuI

These are tetrameric in benzene. A full structural determination on $(Et_3AsCuI)_4$ has shown it to have a structure based on interpenetrating tetrahedra of copper and arsenic atoms (Fig. 8.10).

The silver compounds are similar, but Ph_3PAuCl has been shown to be monomeric.

Complexes of the unipositive metals with oxygen donors are few and unimportant but there are many with sulphur donors. Copper (I) forms thiourea complexes with 1, 2, 3 and 4 thiourea molecules per copper atom; silver (I) complexes are similar. Gold (I) halides form two series of compounds with ethylenethiourea, the salts :

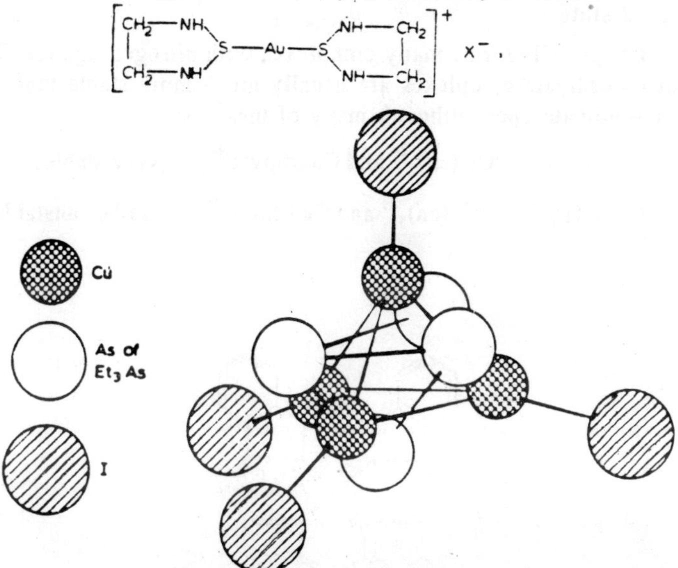

Fig. 8.10. Structure of $[Et_3 AsuI]_4$.

and the covalent compounds :

$$\begin{array}{l} CH_2—NH \\ | \qquad\quad\; \rangle S—Au—X \\ CH_2—NH \end{array}$$

The unipositive metals form cyanocomplexes of empirical formula $M(CN)_2^-$. The silver and gold complexes are monomeric but $KCu(CN)_2$ has been shown to contain spiral, polymeric ions :

$$—Cu—C—N—Cu(—C—N)—C—N—Cu—$$

Copper (I) ammines react with alkynes to give yellow or red polymers in which the copper atom of one RC ≡ CCu unit is π-bonded to the —*C* ≡ *C*— bond of another. Both the ammines and the chlorocuprates (I) absorb carbon monoxide.

The +2 state

Copper (II) forms many complexes with nitrogen ligands. The four-co-ordinate complexes are usually much more stable than the six-co-ordinate ones, although many of these exist :

$Cu(NH_3)_4^{2+}$ $Cu(en)_2^{2+}$ and $Cu(dipy)_2^{2+}$ (very stable)

$Cu(NH)_6^{2+}$ $Cu(en)_3^{2+}$ and $Cu(dipy)_3^{2+}$ (rather unstable)

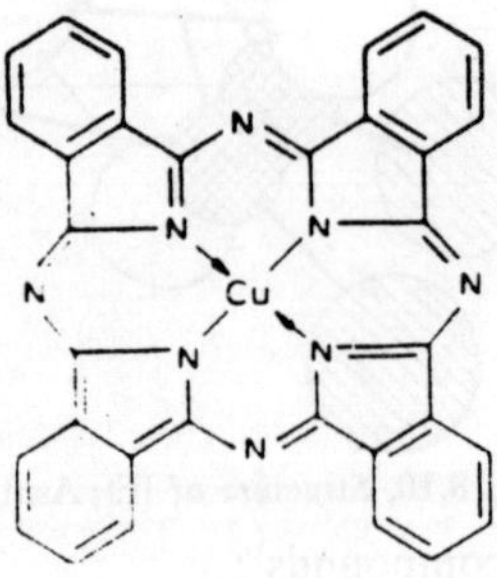

Fig. 8.11. Copper (II) phthalocyanine.

Glycine reacts with Cu^{II} salts to give square planar diglycinecopper, $(NH_2CH_2CO_2)_2Cu$, co-ordinated through both oxygen and nitrogen. There are also numerous complexes with chelating oxygen donors, such as β-diketones, β-ketoesters, catechol and salicylic acid.

Among the copper compounds of technological interest the copper (II) phthalocyanine derivatives are of particular note. The parent compound (Fig. 8.11) is made by heating a Cu^{2+} salt with phthalic anhydride and urea. The derivatives have given colour technologists a series of blue and green pigments of outstanding brightness and colour fastness. Considering the size of the molecule, copper phthalocyanine is remarkably thermally stable; decomposition is not observed below 800 K. The best-known Ag^{II} complexes are those formed by oxidising Ag^+ with persulphate in the presence of nitrogen ligands such as pyridine and bipyridyl : $Ag(py)_4S_2O_8$ is isomorphous with its Cu (II) analogue which is known to have square planar symmetry. Gold (II) complexes are not known.

Effect of solvent on the spectra of copper (II) complexes

Bipositive copper (d^9) can be considered to form square planar complexes in which the odd electron occupies the $d_{x^2-y^2}$ orbital. When additional ligands are available they occupy the fifth and sixth co-ordination positions, but as the d_{z^2} orbital is doubly occupied, these ligands will not approach so closely as those in the plane; the complex thus becomes tetragonal. With the compound in solution the apical positions can be occupied by solvent molecules; the more basic the solvent the stronger the ligand field so created :

The square planar energy diagram for a d^9 system is given on the left of the diagram shown in Fig. 8.12. As solvent molecules approach

38.13

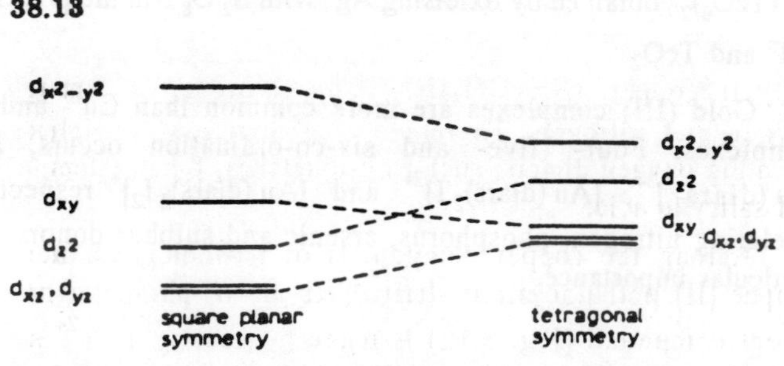

Fig. 8.12. Changes in energy levels as square planar symmetry is converted to tetragonal.

the Cu atom the d_{z^2}, d_{xz} and d_{yz} orbitals increase in energy; at the same time the ligands in the *xy* plane move away from the copper and the $d_{x^2-y^2}$ and d_{xy} energies fall(right of diagram). The extent of this effect depends on the particular solvent.

The absorption spectrum of copper (II) acetylacetonate in a variety of solvents illustrates this (Table 8.4). As the solvent becomes more basic, the $d_{xz} \rightarrow d_{x^2-y^2}$ transition involves a smaller energy change, as does also the $d_{z^2} \rightarrow d_{x^2-y^2}$ transition, but the $d_{xy} \rightarrow d_{x^2-y^2}$ transition is almost unchanged. However, the identification of absorption bands in terms of orbital transitions is rarely as simple as in this d^9 case.

TABLE 8.4 : EFFECT OF SOLVENT ON THE SPECTRUM OF Cu (acac)₂

Wave number/ cm^{-1}	Transition	Dioxan	n-Pentanol	Pyridine	Piperidine
ν_1	d_{xz}, d_{yz} $\rightarrow dx^2-y^2$	17,500	17,100	15,900	15,100
ν_2	$d_{xy} \rightarrow d_{x^2-y^2}$	15,100	15,200	14,800	14,800
ν_3	$d_{z^2} \rightarrow d_{x^2-y^2}$	13,500	13,000	12,100	11,300

The +3 state

Copper (III) is uncommon, but occurs in K_3CuF_6 and in the blue-grey, diamagnetic $KCuO_2$ made by heating CuO with KO_2 in oxygen. Silver (III) occurs in the diamagnetic $KAgF_4$, made by heating $AgNO_3$ and KCl together with fluorine, and in periodate complexes containing $Ag(IO_6)_2^{7-}$ made by boiling Ag_2O with aqueous IO_4^- and O H. There are also complexes containing $Ag(TeO_6)_2^{9-}$ obtained by oxidising Ag^+ with $S_2O_8^{2-}$ in the presence of OH^- and TeO_2.

Gold (III) complexes are more common than Cu^{III} and Ag^{III} complexes. Four-, five- and six-co-ordination occurs, as in $[Au(diars)_2]^{3+}$, $[Au(diars)_2I]^{2+}$ and $[Au(diars)_2I_2]^+$ respectively. Chelating nitrogen, phosphorus, arsenic and sulphur donors are of particular importance.

CHAPTER 9

Analytical and Biological Aspects of Transition Metals

ANALYTICAL ASPECTS

SCANDIUM

Scandium forms a precipitate with 8-hydroxyquinoline (oxine) which has the composition $Sc(C_9H_6ON)_3C_6H_7ON$. The precipitate can be ignited and weighed as Sc_2O_3 for the gravimetric determination of the metal.

Several indicators are available for EDTA titrations of the scandium(III) salt solutions. Erichrome black T can be used in the pH range 7.5-8.00. The colour change at the end point is from red to blue. Another one is mureoxide which can be used in the pH range 2.0-2.6 (acetate buffer). The colour change at the end point is from yellow to violet.

The complexation of scandium with hydroxyflavones has been investigated in detail. Highly sensitive methods involving solvent extraction and ion-exchange techniques have been developed for the estimation of the metal.

TITANIUM

Titanium is precipitated as $TiO_2 . xH_2O$ by the addition of alkali to its salt solution. The precipitate is soluble in dilute acid. An orange-yellow colour develops when the salt solution is treated with H_2O_2. The colour is due to the formation of peroxytitanic oxide $TiO(—O_2—) . xH_2O$.. Fluoride ion discharges this colour due to the formation of colourless hexafluorocomplex $[TiF_6]^{2—}$

Addition of oxalic acid precipitates the white oxalate, soluble in excess of the reagent forming $[Ti(C_2O_4)_3]^{2-}$.

Zinc dust or Sn will reduce Ti(IV)/HCl to violet Ti^{3+} ions. ZR(IV) does not form any coloured ions in similar conditions.

For gravimetric determination, titanium(IV) is precipitated with cupferone. The yellow complex formed is ignitied to the dioxide and weighed. Another method is by precipitation from ≃0.025 N HCl solution by 5,7-dibromo-8-hydroxyquinoline as $[TiO(5,7\text{-oxine})_2]$. It is dried at 120°C and weighed.

VANADIUM

Besides using the coloured vanadyl ion itself several reagents have been used for the detection and photometric determination of vanadium. Methods involving tungstic acid, molybdic acid, hydrogen peroxide, etc. have been found to suffer from lack of sensitivity and selectivity. Organic reagents seem to have an upper edge in respect of selectivity, sensitivity, ease and rapidity of estimation of vanadium ions in various oxidation states.

With borax, vanadium forms a colourless bead in an oxidizing flame which turns green in a reducing flame due to $V(BO_2)_3$.

Acidic vanadate solutions are reduced to VO^{2+} (Blue) ions with mild reducing agents such as SO_2, oxalic acid, Fe^{2+}, etc. Strong reducing agents such as Zn reduce VO_2^+ gradually to blue VO^{2+}, green V^{3+} and finally the violet V^{2+} ions.

The element is carried as vanadate in the soluble portion of Group III of the analysis along with aluminium and chromium. The solution is acidified and treated with lead nitrate when $Pb_3(VO_4)_2$ is precipitated along with $PbCrO_4$. The precipitate is dissolved in 3 N HNO_3 and H_2O_2 added. Blue peroxy chromic acid is extracted in amyl alcohol. A red-brown colour of peroxy vanadic acid appears in the aqueous layer.

Mercurous nitrate precipitates vanadate from weakly acidic solution as Hg_3VO_4. Barium chloride forms a yellow precipitate of $Ba_3(VO_4)_2$.

For gravimetric determinations vanadium is precipitated as Hg_3VO_4 or Ag_3VO-4 and ignited to V_2O_5. Trimetric methods are, however, more convenient. HI in acidic conditions (HCl) reduces vanadates quantitatively to VO^{2+} ; the liberated iodine can be suitably estimated.

CHROMIUM

The element is included in Group III of the element of qualitative analysis and is precipitated as the hydroxide $Cr(OH)_3$. It can be readily oxidized to Na_2CrO_4 with Na_2O_2/alkali for H_2O_2. The chromate is identified by precipitation as yellow $PbCrO_4$ or $BaCrO_4$, in acetic acid solutions. These chromates are soluble in HNO_3. The chromate can also be identified by oxidation with H_2O_2 in presence of acid when blue peroxychromic acid is formed which can be extracted into either or amyl alcohol.

Volumetric estimations of chromium can be carried out by oxidation to dichromate which is then treated with an excess of ferrous ammonium sulphate solution. The excess ferrous is estimated using a standard $K_2Cr_2O_7$ solution.

Chromates give a violet colour on reacting with diphenylcarbizide. The violet coloured complex can be used for the colorimetric determination of chromium (limit of identification 0.5 μg Cr).

For gravimetric determinations, CR(III) is oxidized to chromate and precipitated quantitatively as $BaCrO_4$ or $PBCrO_4$. The precipitate is dried and weighed as such.

MANGANESE

The element is detected in Group IV of the scheme of analysis, where it is precipitated as buff coloured MnS in NH_4Cl-NH_4OH medium. Mn^{2+} is confirmed by oxidation to the violet MnO_4^- ion. The oxidation can be effected with PbO_2 in conc. HNO_3 or with sodium bismuthate in cold conc. HNO_3.

For gravimetric determination, Mn^{2+} can be precipitated from ammoniacal solutions as $MnNH_4PO_4 \cdot 7H_2O$ and ignited to $Mn_2P_2O_7$ and weighed.

Volumetric estimations of Mn(I) can be carried out by oxidation of the salt solutions to MnO_4^- ; addition of excess of standard Fe(II) and determination of the excess by titrating against standard $KMnO_4$ solution.

Mm(II) can also be estimated by direct complexometric titrations against EDTA at pH 7 using triethanolamine as buffer and hydroxylamine to keep manganese in the divalent state.

IRON

Iron(III) is precipitated as the hydroxide $Fe(OH)_3 . xH_2O$ in the Group III of the scheme of analysis. It is detected by the red colour with thiocyanate solution or by the formation of 'Prussian blue' with ferrocyanide.

Iron(II) solutions from deep red water soluble complexes with α, α'-dipyridyl, *o*-phenanthroline or dimethlyglyoxime. A brown black complex $[Fe(H_2O)_5NO]^{2+}$ is formed when NO_2^- is added to an acidic solution of Fe(II). This forms the basis for the brown ring test for NO_2^- and NO_3^-.

Iron(III) is estimated volumetrically by reducing it to iron(II) by excess stannous chloride (the excess being removed with $HgCl_2$), and titrating the ferrous iron with $KMnO_4$ or $K_2Cr_2O_7$. $KMnO_4$ acts as self-indicator while in case of $K_2C_2O_7$ diphenylamine is used as the indicator.

COBALT

Cobalt appears in the Group IV of the scheme of analysis, where it is precipitated as CoS (black) along with the sulphides of Ni, Zn and Mn. They are precipitated by passing H_2S through ammoniacal solution. Cobalt is detected by formation of a blue colour with thiocyanate $[Co(SCN)_4]^{2-}$. The complex a be extracted into amyl alcohol. It can also be confirmed by the appearance of a yellow precipitate when warmed with KNO_2 due to the formation of $K_3(CO(NO_2)_6]$.

For volumetric estimations cobalt(II) can be directly titrated against EDTA in the presence of hexamine using xylenol orange as indicator.

Cobalt(II) forms a red-brown complex with 1-nitroso-2-naphthol in carbon tetrachloride. This reaction is used for the colorimetric determination of cobalt.

NICKEL

Nickel(II) appears in Group IV of the scheme of analysis as NiS. It is detected and also quantitatively estimated as the red coloured bis(dimethylglyoximato)nickel(II) in slightly ammoniacal medium. Small amounts of cobalt are also coprecipitated $Ni(DMG)_2$ after extracting with $CHCl_3$ can be used for colorimetric determination of nickel(II).

COPPER

Copper(II) is precipitated as CuS from acidic (HCl) solutions in Group(II) of the scheme of analysis. CuS is soluble in hot 50 percent HNO_3. In the presence of KCN, CuS is not precipitated due to the formation of $[Cu(CN)_3]^{2-}$ ions.

A blue precipitate of $Cu(OH)_2$ is obtained with NaOH. The precipitate dissolves in ammonia forming $[Cu(NH_3)_4]^{2+}$.

For quantitative estimations copper(II) is reduced to copper(I) by SO_2 in dilute acidic medium and precipitated as CuSCN by adding thiocyanate. The precipitate is dried and weighed as such.

From neutral copper(II) solutions, KI forms unstable CuI_2 which decomposes quantitatively into CuI and I_2. The liberated iodine can be estimated volumetrically by titrating against a standard thiosulphate solution.

ZINC

All zinc compounds, on fusion with Na_2CO_3 in a charcoal cavity give ZnO which is yellow when hot and turns white on cooling.

Zinc(II) is precipitated as ZnS by passing H_2S in ammoniacal medium. The precipitate is soluble in dilute (2 N) HCl. With alkali, Zn(II) forms $Zn(OH)_2$ which dissolves in excess of alkali forming zincates. Zn(II) is detected as the insoluble dirty white $K_2Zn_2[Fe(CN)_6]_2$.

For quantitative determination Zinc is precipitated as $ZnNH_4PO_4$ at pH 6-7, dried at 150°C and weighed as such for ignited and weighed as $Zn_2P_2O_7$.

Zinc(II) ions can be titrated against EDTA at pH 10.0 using erichrome black T as indicator.

BIOLOGICAL ASPECTS

Among the transition metal ions, iron, copper, manganese, cobalt, molybdenum and zinc play important roles in the biochemistry of the human body. Organomercury compounds have been used for centuries as pharmaceuticals. Excess of these ions are, however toxic.

Most of these ions possess three chemical properties that are extremely important in carrying out biochemical processes. First they form stable complexes with proteins and other biologically active substances ; they are very good catalysts, and finally they exhibit various oxidation states which are readily formed and interconverted in the environment of the cell. The interconversion of oxidation states is uniquely suited to biochemical functions.

One important function of these metal ions in their role as a *cofactor* is reactions involving enzymes. Cofactors are non-protein constituents of enzymes which must be bound to the enzyme in order for it to be catalytically active. Some enzymes require cofactors such as metal ions or coenzymes. The metal ions acting as cofactors in such enzymes may act either by coordinating to the reactant species and binding it to the enzyme or they mat act as the catalyst for the reaction itself. Some enzymes that need metal ions as cofactors are given in Table 9.1.

TABLE 9.1 : SOME ENZYMES THAT REQUIRE COFACTORS.

Cofactor	Enzyme	Reaction
Zn^{2+}	Carboxypeptidase	Hydrolysis of proteins
Cu^{2+}	Ascorbic acid oxidase	Oxidation of ascorbic acid
Mn^{2+}	ATPase	Transfer of phosphate
CO^{2+}	Cobamide	Synthesis of amino acids

IRON

Iron as a vital constituent of every mammalian cell. The role of iron in the body is closely associated with haemoglobin and the transport of oxygen from the lungs to the tissue cells.

An average human body weighing 70 kg has on estimated iron content of about 5 g. All of it is bound to proteins in one form or another. Blood haemoglobin represents about 60-70% of total iron in the body, i.e., about 3-3.5g. This level is maintained by absorbing merely 1 mg of iron per day. Myoglobin contains about 3-5% of body iron and haem-enzymes about 0.2% of body iron.

The non-haem compounds contain about 15% of the total body iron. These proteins transfer and store the iron itself.

The haem proteins transport oxygen and electrons, whereas the main function of the non-haem proteins is transfer and storage of iron itself.

In the haemoglobin molecule, an atom of ferrous iron, with a coordination valence of six, exists at the centre of each four tetrapyrrole rings. Four of the valences bind the pyrrole nitrogens in the plane of the ring while in the fifth above or below and perpendicular to the plane, is attached to the nitrogen atom of a protein. This configuration permits the reversible combination with molecular oxygen (depending upon the partial pressure) at the sixth coordination position of iron. Thus, the uptake of oxygen is facilitated in the lungs where the partial pressure is high, and is released in the tissues where the partial pressure of oxygen is low.

Unfortunately the affinity of iron for CO is over 200 times that of oxygen. Thus, it acts as a fatal poison. The ability of Cn^- ions to replace oxygen is the reason for its poisonous nature.

Ferritin, a non-haem protein is considered to be essentially an iron storage protein. It transports iron to the blood whenever required. In human body, the liver and the spleen are the main storage organs for iron in the form of ferritin. The nutritional requirement for iron in the form of ferritin. The nutritional requirement for iron is exceedingly small and vary with age and sex. The human adult male absorbs less than 5 mg of iron per day. Women absorb slightly more. Dietary iron should be taken in the ferrous state. Much of iron in the food is in the ferric form. A part of iron is of by the gastric HCl and subsequently reduced by reducing agents present in food such as ascorbic acid.

Among the foods rich in iron are the meats—liver, fish and egg yolks— and vegetables—especially leafy green vegetables, peas, beans, etc. Milk and fresh fruit are poor in iron.

MANGANESE AND COBALT

The human body requires about 12 to 20 mg of manganese. The highest concentrations of the metal are present in the liver, kidney and the pancreas of human beings.

Manganese(II) complexes with ADP or ATP to facilitate the transfer of phosphate, a process that is vital in the energy-utilization cycle.

Cobalt ion is present in cobamide, a derivative of vitamin B_{12}, and one of the most extraordinary of all biologically active substances. Cobamide is important in the *in vivo* (in the body) synthesis of amino acids used to make proteins. Deficiency of cobalt can lead to pernicious anemia.

COPPER

The role of copper in human biochemistry is especially interesting. Normally, abundance of copper is supplied in our water and food and the copper ware we use in cooking.

Copper deficiencies lead to degeneration of the sheath around the spinal cord, weakening of the walls of certain blood vessels (including the aorta), decolouration of hair, reduction in the synthesis of haemoglobin and phospholipids and inhibition of the energy-producing reactions of the cell.

The human adult requirement of copper is 2 mg per day, and the adult human body contains about 100-150 mg of copper, the greatest concentration existing in the liver and bones. Blood contains a number of copper proteins, and copper is known to be necessary for the synthesis of haemoglobin, although there is no copper present in the haemoglobin molecule.

If a radioactive copper salt is injected into the human body, the copper ion first appears in combination with the serum albumin in the blood serum. It is then rapidly absorbed by the liver and later appears in serum as the blue copper-containing protein called ceruloplasmin. This protein serves as a reservoir for copper and as a catalyst for certain reactions. Ceruloplasmin supplies copper ion to certain enzyme systems in the tissues.

Copper is a constituent of some of the cytochrome enzymes, which are catalysts for the main respiratory reaction chains, involving transfer of electrons from various carbohydrates to oxygen. The enzyme tyrosinase is present in many animals including humans. It is mainly responsible for skin pigmentation and controls hair colour.

The blue copper protein haemocyanin occurs in the blood of lower forms of animal life (snails, crabs, scorpions and octopuses). This protein, believed to be a polypeptide containing copper(I), performs the oxygen-carrying function for these species. It is not, however, as efficient an oxygen carrier as haemoglobin.

If ceruloplasmin in the serum were destroyed, the copper ions would diffuse into the tissues where they could accumulate at a high level in the liver and in the brain. This is known as ***Wilson's disease*** and can produce several mental illness and even death. However, this condition can be treated by reducing copper in the diet by eliminating foods such as nuts, mushrooms, liver and oysters. Another alternative is by using a chelating drug such as penicillamine, which leaches the copper out of the tissues.

Ingestion of copper sulphate by human causes vomiting, cramps, convulsions, etc. and as little as 27 g of the compound may cause death. The toxicity of copper is probably partly due to its combination with thiol groups of certain enzymes, thereby inactivating them.

Copper is also essential in plant metabolism, where it is active in the synthesis of chlorophyll and several reactions involving enzymes. Several enzymes catalysing oxidation reduction reactions (oxidases) are known to contain copper. The activity of copper in these enzymes is believed to be due to its interconversion between +1 and +2 oxidation states.

Ascorbic acid oxidase catalyses the reaction between oxygen and ascorbic acid to form dehydroascorbic acid.

Tyrosinase occurs in potatoes, spinach, mushrooms, and other plants. It catalyses the oxidation of monophenols to diphenols and catechol to melanins.

Traces of copper are required for the growth and reproduction of lower plant forms such as algae and fungi. Copper deficiency in plants causes chlorosis, reduced photosynthetic activity, and inability to produce seed. Excess of copper in the soil is toxic.

ZINC

Zinc is biologically one of the most important metals and is apparently necessary to all forms of life.

$$\text{--NH--CH(R)--C(=O)--NH--CH}(\text{CH}_2\text{--C}_6\text{H}_4\text{--OH})\text{--CO}_2^- \xrightarrow[\text{H}_2\text{O}]{\text{enzyme}} \text{--NH--CH(R)--CO}_2^- + \text{H}_3\text{N}^+\text{--CH}(\text{CH}_2\text{--C}_6\text{H}_4\text{--OH})\text{--CO}_2^-$$

An adult human body contains about 2 mg of zinc. *Carboxy-peptidase* A and *carbonic anhydrase* are the two enzymes which have received most attention.

Carboxypeptidase A catalyses the hydrolysis of proteins to amino acids which occur in the small intestines during the process of digestion.

It has a molecular weight of about 30,000 to 35,000 and contains one zinc tetrahedrally coordinated to 2 histidine N atoms, a carboxyl O of a glutamate residue, and a water molecule.

Carbonic anhydrase was the first zinc metalloenzyme to be discovered. Three closely related forms A, B and C are found in mammalian erythrocytes (red-blood cells), where they catalyse the reaction :

$$CO_2 + H_2O \rightleftharpoons HCO_4^- + H^+$$

The forward reaction takes place when CO_2 is taken up by blood in tissue, while the backward reaction occurs when CO_2 is released in the lungs. The enzyme increases the reaction rate by about a million times.

CHAPTER 10

Coordination Compounds

INTRODUCTION

In the middle of the eighteenth century, an artist's pigment known as Prussian blue was first prepared when sodium carbonate and animal excrement were heated together in an iron pot. It is now known that Prussian blue is a *coordination compound* which may be prepared by the action of iron(III) ion with $K_4Fe(CN)_6$. Early workers termed substances such as Prussian blue "Complex compounds," and this name has persisted. Even today the term *complex compound* is used synonymously with the term *coordination compound.*

A large variety of chemical substances may be classified as coordination compounds. In aqueous solution, all metal ions acquire water of hydration and become complex ions. Coordination compounds play an essential role in the chemical industry and life itself.

The well known Ziegler-Natta catalyst, used for the low-pressure polymerization of ethylene is a complex of aluminium and titanium. Complex formation is used extensively in analytical chemistry for detection, estimation and separation of metal ions.

In living creatures, two important coordination compounds are chlorophyll in green plants and hemoglobin in the blood of vertebrate animals. Chlorophyll, which is vital to photosynthesis in plants, is a magnesium complex and hemoglobin, which carries, oxygen to animals cells, is an iron complex. Structures of the two complexes are shown in Fig.10.1

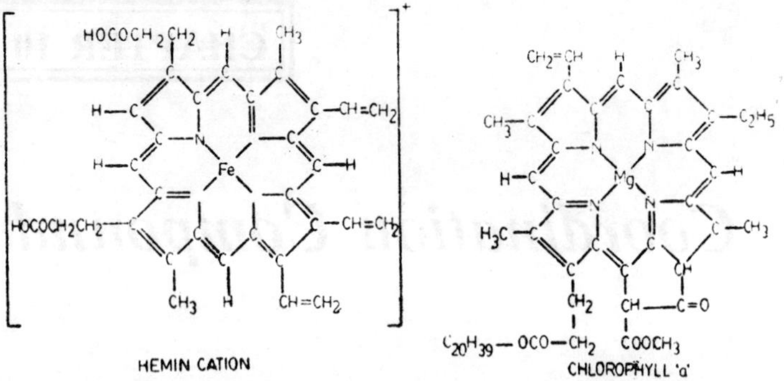

HEMIN CATION

CHLOROPHYLL 'a'

Fig. 10.1. Structures of chlorophyll and hemoglobin.

DEFINITIONS

A coordination compound may be defined as a compound containing a central atom or ion to which are attached molecules or ions whose number usually exceeds the number corresponding to the oxidation number or valence of the central atom or ion. The *central atom*, usually a transition metal, acts as a Lewis acid.

The molecules or ions which attach to the central metal ion are called *ligands*. They are electron pair donors and thus act as Lewis bases. Ligands may be neutral molecules such as ammonia, water, ethylenediamine, etc. They may also be ions such as Cl^-,CN^-,NO_2^-,SCN^-, etc.

Ligands may be unidentate, that is, they may possess only one coordinating atom. Examples of unidentate ligands are NH_3,H_2O,Cl^-,CN^-, etc. Bidentate ligands may possess two coordinating atoms. Ethylenediamine and oxalate are examples of bidentate ligands. When a bidentate, tridentate or a polydentate ligand forms coordinate bonds with the same metal ion, the ring structure formed is called a *chelate ring* and the ligand is called a chelating agent. Some of the most common ligands are shown below along with their structures. The name of the ligand is given below its structure and the symbol is given in parentheses.

Monodentate Ligands

Cl^-	Br^-	I^-	CN^-
Chloride	Bromide	Iodide	Cyanide
(chloro)	(bromo)	(iodo)	(cyano)

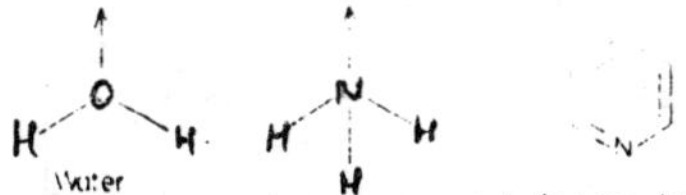

Water (aqua) Ammonia (ammine) Pyridine (py)

Bidentate Ligands

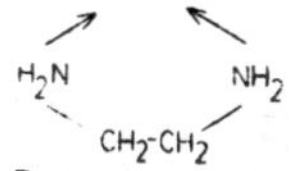

Ethylenediamine (en)

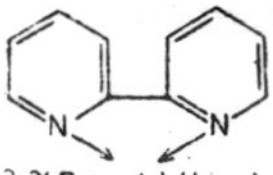

2,2'-Bipyridyl (bipy)

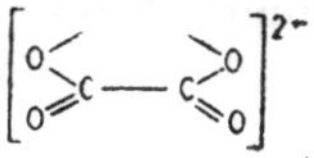

Oxalate ion (ox)

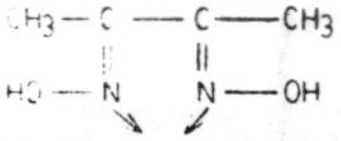

Dimethylglyoxime (DMG)

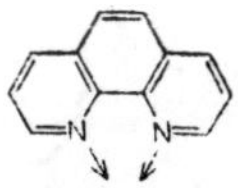

1,10 Phenanthroline (phen)

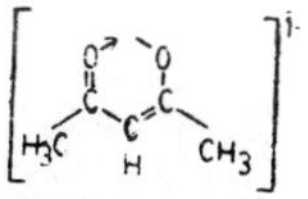

Acetylacetonate ion (acac)

Tri- OR Terdentate Ligands

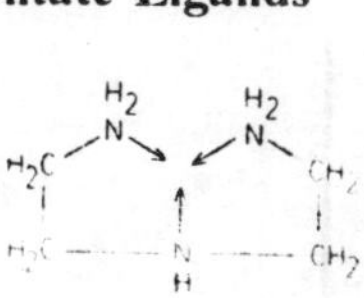

DIETHYLENETRIAMINE (dien)

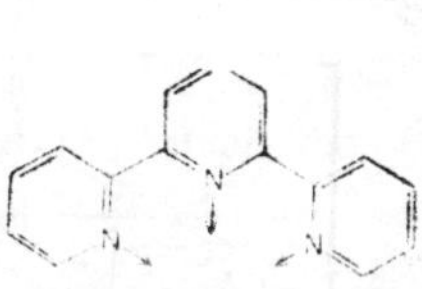

2,2',2''-TERPYRIDINE (terpy)

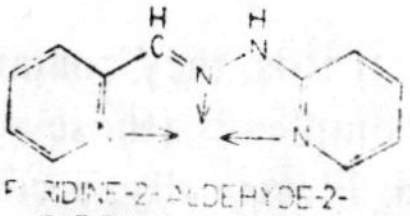

PYRIDINE-2-ALDEHYDE-2-PYRIDYLHYDRAZONE

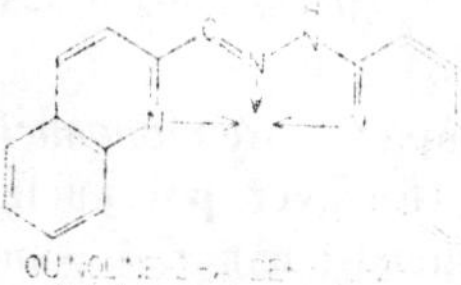

QUINOLINE-2-ALDEHYDE-2-QUINOLYLHYDRAZONE

Tetra- OR Quadridentate Ligands

TRIETHYLENETETRAMINE (trien)

ETHYLENEBIS(SALICYLALDIMINE) ION

Penta- OR Quinquedentate Ligands

ETHYLENEDIAMINETRIACETATE ION

Hexa- OR Sexadentate Ligands

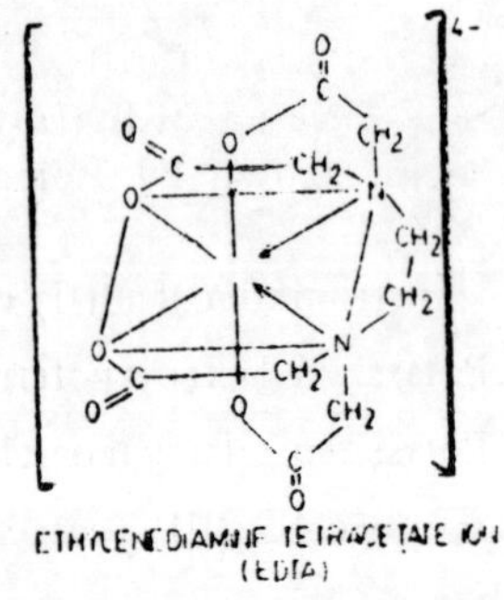

ETHYLENEDIAMINE TETRACETATE ION (EDTA)

Most complexes are *mononuclear*, that is, they contain only one central atom. However, polynuclear complexes (those with two or more central metal ions) are also known. Electrically, a complex may be positive, negative or even neutral. The charge on a complex simply depends upon the balance between the charges of the central atom and the ligands. The ligands together with the central atom constitute the *coordination sphere.* The bonding between the central atom and the ligands possesses substantial covalent character. Although partial dissociation may occur in some cases, the complex usually tends to remain as a discrete unit, even in solution.

The total number of ligands surrounding and bonded to a central atom is known as the *coordination number* of the central atom. Coordination numbers from one to ten are known, but the most common are four and six. The coordination number of silver ion in the complex $[Ag(N_3)_2]^+$ is two and that for iron in $[FeCl_4]^-$ and $[Fe(CN)_6]^{3-}$ are four and six respectively.

NOMENCLATURE OF COMPLEX COMPOUNDS

Coordination compounds are named according to the following detailed set of rules, which have been adopted by the International Union of Pure and Applied Chemistry, IUPAC (1976) :

1. While naming a salt as well as a complex, the cation is named *first* followed by the anion.

   KCl Potassium chloride

   $K_2[PtCl_6]$ Potassium hexachloroplatinate

2. If the complex is a cation or a neutral molecule, the name of the central metal atom is used unchanged, followed immediately by Roman numeral in parentheses to indicate its oxidation state, e.g. nicke(II), copper(II), cobalt(III), etc. When the complex is anionic, the suffix *ate* is added to the name of the metal, e.g. ferrate, cobaltate, etc. The formal oxidation state is again shown by Roman numerals in parentheses :

   $[Co(NH_3)_6]Cl_2$ Hexaammincobalt(II) chloride

   $K_4[Fe(CN)_6]$ Potassium hexacyanoferrate(II)

   $K_4[Ni(CN)_4]$ Potassium tetracyanonickelate(O)

   $[Co(NH_3)_6][CuCl_5]$ Hexaamminecobalt(III) pentachlorocuprate(II)

3. Names of neutral ligands are used unchanged, except for H_2O, which is called aqua ; —NH_3, which is called ammine ; —CO, which is called carbonyl ; and —NO which is called nitrosyl.

4. The names of negative ligands end in the suffix *-o* and positive ligands end in —*ium.*

For example :

Negative ligands	*Positive ligands*
Cl^- chloro	NO_2^+ nitronium
Br^- bromo	
I^- iodo	NH_2
	|
CN^- cyano	NH_3^+ hydrazinium
SO_4^{2-} sulphato	$C_5H_5N^+$ pyridinium
$C_2O_4^{2-}$ oxalato	
NO_3^- nitrato	
NH_2^- amido	

5. The constituents of a coordination sphere are listed as follows :

Ligands are named first followed by the central metal atom at the end.

The ligands are listed in *alphabetical order* regardless of the number of each. The name of the ligand is treated as a unit, and the prefix di— , tri — , etc. are not considered. Thus, "diammine" is listed under "a" and "dimethylammine" under "d". Examples are :

$[Cr(H_2O)_5Cl]^{2+}$ pentaaquachlorochromium(III) ion

$[Pt(NH_3)_2Cl_2]$ diamminedichloroplatinium(II)

$[Co(NH_3)_4Cl_2]Cl$ tetraamminedichlorocobalt(II) chloride

6. The prefixes di—, tri—, etc. are used to indicate the number of ligands of one type. For complex ligands which already have di—, tri—, etc. in their names, the prefixes bis, tris, tetrakis, etc. are used. For example, $[Cu(en)_2]^{2+}$ is named as bis (ethylenediammine) copper(II) ion.

7. A bridging group is indicated by adding the Greek letter μ, immediately before its name and separating the name from the rest of the complex by hyphens.

Two or more bridging groups of the same type are indicated by di μ⁻ (or bis⁻ μ⁻), etc.

The bridging groups are listed with the other groups in alphabetical order. Where the same group is present as a bridging ligand and as a simple ligand, it is cited first as a bridging ligand.

Examples :

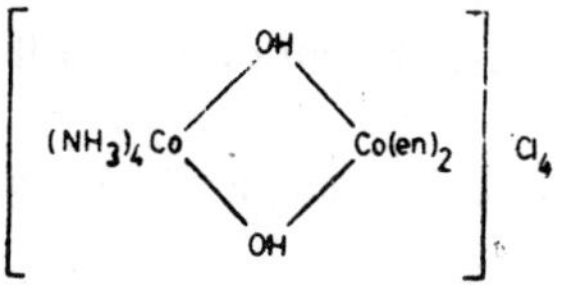

OH
(NH3)4 Co Co(en)2 Cl4
OH

Tetrammine- μ -dihydroxobis(ethylenediamine)dicobalt(III) chloride

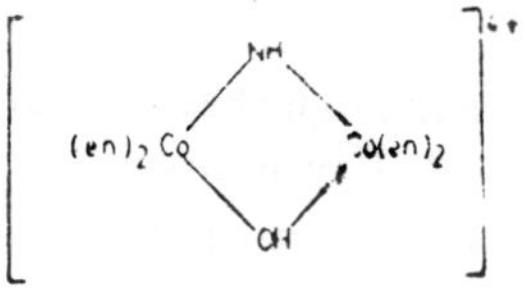

NH
(en)2 Co Co(en)2
OH

Tetrakis(ethylenediamine)-μ -hydroxodicobalt(III) ion

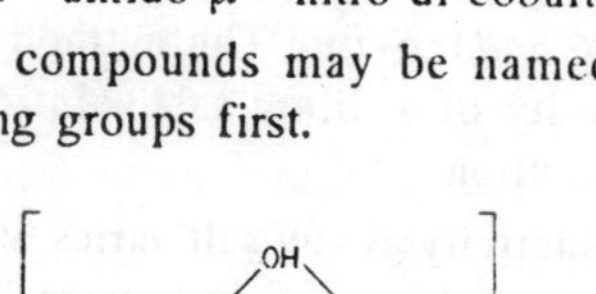

NH2
NH3)4 Co Co(NH3)4 (NO3)4
NO2

Octammine-μ - amido-μ - nitro-di-cobalt(III) nitrate

Symmetrical compounds may be named more compactly by placing the bridging groups first.

Examples :

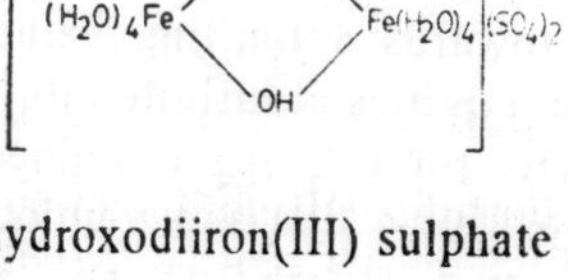

OH
(H2O)4 Fe Fe(H2O)4 (SO4)2
OH

Octaqua-μ-dihydroxodiiron(III) sulphate

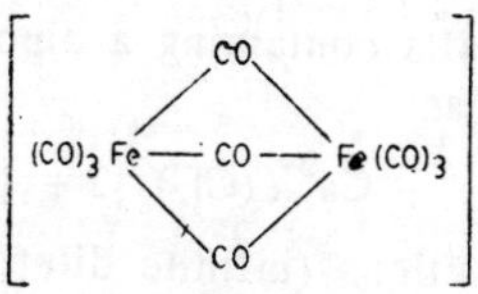

CO
(CO)3 Fe — CO — Fe (CO)3
CO

Tri-μ-carbonylbis(tricarbonylir

8. In case of a ligand having more than one donor atom, its point of attachment with the central metal atom may be denoted by adding the italicized symbol for the atom, through which attachment occurs, at the end of the name of the ligand.

Thus the dithiooxalato anion which may be attached through

$$\begin{array}{c} {}^{-}S \qquad\qquad S^{-} \\ \diagdown \qquad \diagup \\ C - C \\ /\!/ \qquad \backslash\!\backslash \\ O \qquad\qquad O \end{array}$$

S or O are distinguished as dithiooxalato—S,S′ and dithiooxalato—O,O′ respectively.

9. Geometrical isomers are named as *CIS*- or trans- depending upon whether the similar ligands are on the same side or otherwise.

DETECTION OF COMPLEX COMPOUNDS

The best evidence of the formation of a complex is its isolation and determination of the definite composition. This, however, is not always possible. Hence, other methods have to be adopted for the detection of complex formation. These methods depend upon the changes in physical or chemical properties of the species. Some such methods are given below.

1. *Changes in Conductivity* : The method is based on the fact that molar conductivity of a dilute salt solution depends upon the number of ions in solution.

The molar conductivity of any salt varies with concentration, but as the solutions are made more and more dilute, the observed conductivities tend towards a limiting value. The magnitude of observed molar conductivities at infinite dilution depends upon the charge type of the electrolyte being examined. For example, salts consisting of a unipositive cation and a unipositive anion, such as K^+Cl^-, [represented as (1 + , 1 –)] have a lower molar conductivity at infinite dilution than salts containing a dipositive cation and two unipositive anions such as

$$Ca^{2+}, (Cl^-)_2\,[2+, 1-].$$

In very dilute solutions (infinite dilution) the nature of the constituents of the salt has little effect on the molar conductivity of the salt. Thus a sudden appreciable change in the molar conductivity of the compound indicates complex formation.

2. *Changes in pH Values* : When the complex formation involves change in H^+ or OH^- ion concentration, the pH change of the solution can be used as a means of detection of complex formation. For example, in the formation of metal complexes with glycine, oxalate, EDTA, etc. the H^+ ion concentration increases and hence the pH decreases.

$$Fe^{3+} + 3H_2C_2O_4 \rightarrow [Fe(C_2O_4)_3]^{3-} + 6H^+$$

3. *Polarographic Method* : Many metal ions are reducible at the dropping mercury electrode and have characteristic half-wave potentials, $E_{1/2}$. A shift of $E_{1/2}$ towards more negative values indicates complex formation.

4. *Potentiometric Method* : The method depends upon the concentration of the species in solution. As complex formation takes place, the concentration of the metal ion M^{n+} decreases. The change in the electrode potential $E°$ 1, of the system $M^{n+} - M$ vs the concentration of the ligand L, is followed to study the complexation equilibrium.

5. *Changes in Solubility* : Certain complexes are less soluble in aqueous solution and are precipitated on formation, for example $[Al(acac)_3]$ and $[Ni(DMG)_2]$, etc. Neutral chelates of organic ligands generally dissolve in non-polar solvents like ether, benzene, chloroform, etc.

Sometimes, a precipitate may dissolve in the solution of a complexing agent. For example, AgCl is insoluble in aqueous solution but dissolves in ammonia forming $[Ag(NH_3)_2]^+$ ion ; HgI_2 similarly dissolves in KI solution forming $HgI_4]^{2-}$ ion.

6. *Colour Change* : A change in colour is a definite indication of complex formation. Some such cases are :

$$\underset{\text{(pink)}}{Co^{2+}} + 4Cl^- \rightarrow \underset{\text{(blue)}}{[CoCl_4]^{2-}}$$

$$\underset{\text{(yellow)}}{Fe^{3+}} + nSCN^- \rightarrow \underset{\text{(red)}_n}{Fe(SCN)]^{3-n}}$$

$$\underset{\text{(white)}}{Ni(CN)_2} + 2KCN \rightarrow \underset{\text{(orange-red)}}{K_2[Ni(CN)_4]}$$

7. *Chemical Reactions* : Standard chemical reactions can be used to detect complex formation. Thus metal hydroxides cannot be precipitated from metal salt solutions in the presence of EDTA (formation of metal EDTA complex). H_2S cannot precipitate CuS from a solution of Cu^{2+} ion in the presence of CN^- ions due to the formation of $[Cu(CN)_4]^{2-}$ complex.

ISOMERISM IN COORDINATION COMPLEXES

In addition to stereoisomerism, which arises due to different spatial arrangement of ligands within the coordination sphere, coordination complexes exhibit the following types of isomerism.

Ionization Isomerism

The salts having identical composition which produce different ions when dissolved in water are known as ionization isomers. The isomerism arises due to the exchange of groups between the coordination sphere and ionization sphere. Examples are :

$$[Co(NH_3)_5SO_4]Cl \quad \text{and} \quad [Co(NH_3)_5Cl]\,SO_4$$

The first compound gives immediately a precipitate of AgCl with $AgNO_3$ but does not give a precipitate of $BaSO_4$ upon treatment with a barium salt. The other isomer behaves in the opposite manner.

Linkage Isomerism

Linkage isomers are different compounds which contain a ligand capable of attachment to the central metal ion through either of two different donor atoms. For example, the nitrite ion, NO_2^-, may attach to the central metal ion by an oxygen atom or by the nitrogen atom. Thus, the ions

$$[Co(NH_3)_5NO_2]^{2+} \text{ and } [Co(NH_3)_5ONO]^{2+}$$

can exist as isomers. Another ligand which could exhibit linkage isomerism is the thiocyanate ion SCN^-. The ligand can coordinate to the metal ion through either the sulphur atom or the nitrogen atom. Linkage isomers containing the SCN^- ion can be distinguished by infrared spectroscopy. If the SCN^- ion coordinates to the metal through the S atom, the frequency of the C-S stretching vibration is lower than that observed for the free ion. On the other hand, if coordination is through nitrogen, the frequency of the C-S stretching vibration is higher than that for the free ion.

Coordination Isomerism

When a salt contains complex cations and anions, it may exhibit isomerism through the interchange of ligands from cation and anion. This type of isomerism is called coordination isomerism and is illustrated by the following examples :

$$[Co(NH_3)_6]\,[Cr(CN)_6]$$

and its coordination isomer

$$[Co(CN)_6]\,[Cr(NH_3)_6]\ ;$$

$$[Co(NH_3)_6]\,[Cr(NO_2)_6]$$

and its isomer

$$[Co(NO_2)_6]\,[Cr(NH_3)_6]\ ;$$

$$[Cu(NH_3)_4]\,[PtCl_4]$$

and $$[(CuCl]_4\,[PtCNH_3)_4]$$

A special type of coordination isomerism, sometimes known as *coordination position isomerism*, involves the different placement of ligands in a complex. Examples of this type include

$$\left((NH_3)_4\,Co\overset{H}{\underset{H}{\overset{O}{\underset{O}{\lessgtr}}}}Co(NH_3)_2Cl_2)\right)^{2+}$$

and

$$\left(Cl(NH_3)_3\,Co\overset{H}{\underset{H}{\overset{O}{\underset{O}{\lessgtr}}}}Co(NH_3)_3Cl)\right)^{2+}$$

Polymerization Isomerism

Complexes having the same empirical composition but different molecular formulae were classified by Werner as *polymerization isomers.* Thus $[Pt(NH_3)_2\,Cl_2]$ and $[Pt(NH_3)_4]\,[PtCl_4]$ are polymerization isomers. The former is a non-electrolyte and the latter conducts as (2 + , 2—) electrolyte. The use of the term polymerization is however, improper since polymerization refers to building a larger structure consisting of the same repeating units.

STEREOISOMERISM

Two types of stereoisomerism are exhibited by the coordination compounds, *geometrical isomerism* and *optical isomerism.*

Werner's theory provided for the existence of geometric isomers. If the coordination sphere has a definite geometry, some sets of ligands can occupy the sites in the coordination sphere in more than one way. This type of isomerism is often referred to as *cistrans-isomerism.* Such isomerism is not possible for complexes having coordination number 2 or 3 and also for tetrahedral complexes (coordination number 4). This is due to the fact that in these arrangements all the positions are adjacent to each other. For example, in tetrahedral zinc(II) complex $[Zn(NH_3)_4]^{2+}$ if two $—NH_3$ were replaced by two Cl^- ions, the resulting compound $Zn(NH_3)_2Cl_2$, would exist in only one form with no isomers. In contrast a square planar complex $Pt(NH_3)_2Cl_2$ can exist in *cis-* and *trans-* isomeric forms. Octahedral complexes also exhibit this type of isomerism. Some examples are represented below :

Square planar complexes. $[Mx_2y_2]$ type of square planar complexes, where x or y is a monodentate ligand (Fig. 10.2).

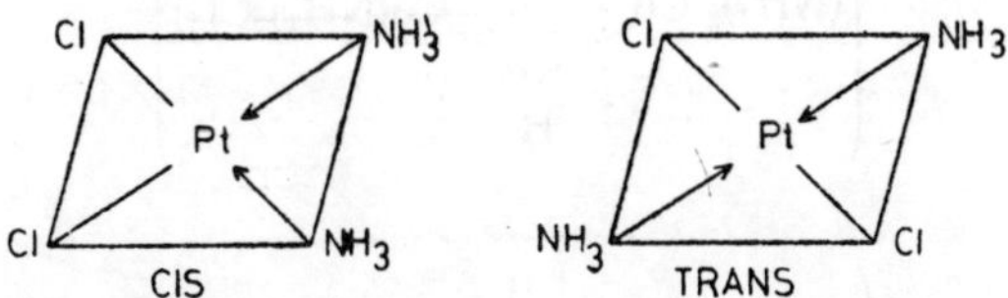

Fig. 10.2. Geometrical (*cis-* and *trans-*) isomers of dichlorodiammine-platinum(II).

Square planar complexes having two *unsymmetrical bidentate ligands* (Fig. 10.3).

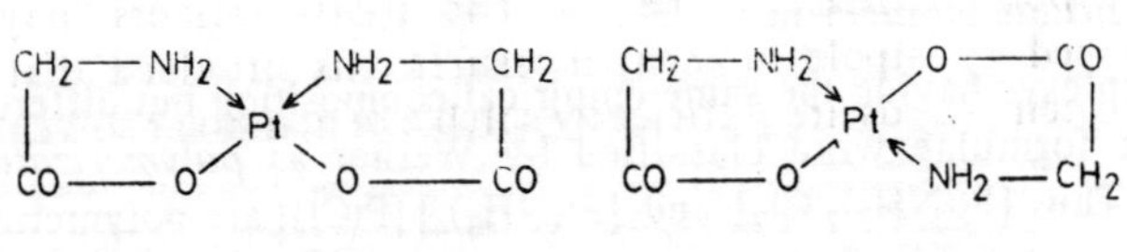

Fig. 10.3. Geometrical (*cis-* and *trans-*) isomers of diglycinatoplatinum(II).

Octahedral complexes. For monodentate ligands x and y, no isomers are possible for $[Mxy_5]$ or $[Mx_5 - y]$.

$[Mx_4y_2]$ *type of octahedral complexs.* In *cis-* isomer the two monodentate *y* ligands are in 1, 2-positions while in *trans-* isomer they are in 1, 6-positions (Fig. 10.4).

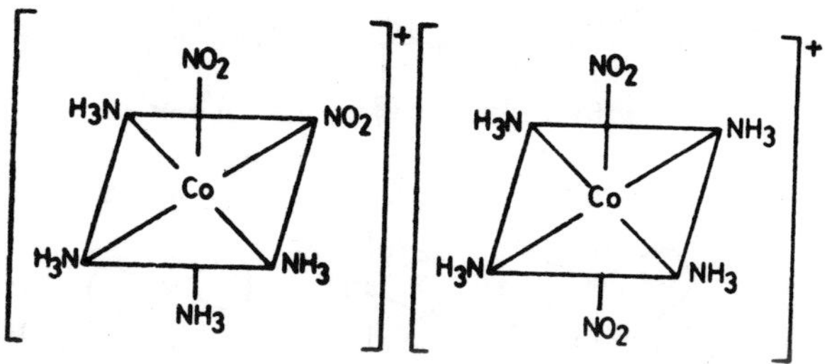

Fig. 10.4. Geometrical (*cis-* and *trans-*) isomers of dinitrotetramminecobalt (III) ion.

$[Mx_3y_3]$ type of octahedral complexes (Fig 10.5).

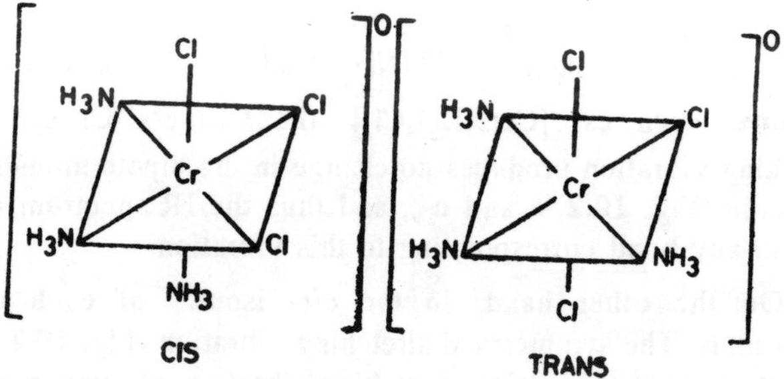

Fig. 10.5. Geometrical (*cis-* and *trans-*) isomers of trichlorotriamminechromium (III).

A bidentate ligand (*x y*) can span only two *cis-* positions. Hence, $[M(xy)A_3B]$, $[M(xy)A_2B_2]$ and $[M(xy)_2A_2)$ exhibit geometrical isomerism. These isomers are named according to the arrangement of the A and B groups (Fig. 10.6).

This *cis-* and *trans-* isomers can be distinguished by :

1. *Dipole Moments* : The *cis-* and *trans-* isomers may be distinguished by dipole moment measurements provided that the respective compounds are sufficiently soluble in non-polar solvents to enable measurements of their dipole moments. For compounds of the type $[Pt\{P(C_2H_5)_3\}_2Cl_2]$, it has been shown that the trans isomers have a dipole moment of zero, whereas the *cis-* isomers have dipole moment values of 8 to 12 Debyes.

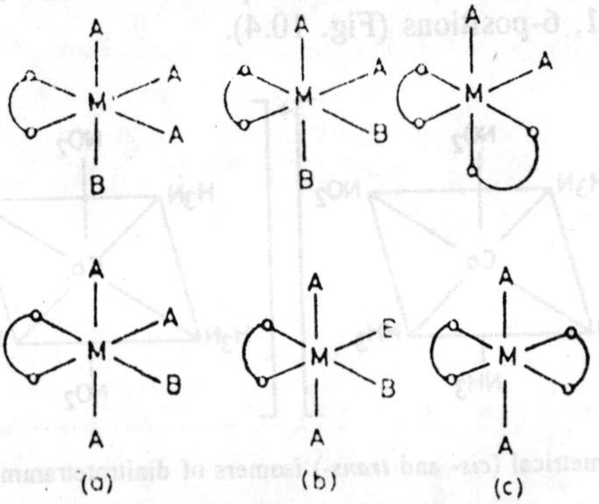

Fig. 10.6. *cis-* and *trans-* isomers of (a) $M(xy)A_3B$, (b) $M(xy)A_2B_2$ and (c) $M(xy)_2A_2$

2. *Infrared Spectroscopy* : IR spectroscopy is useful in distinguishing between *cis-* and *trans-* isomers. In a *trans-* square planar complex, such as $[Pt(NH_3)_2Cl_2]$, or a *trans-* octahedral complex, such as, $[Co(NH_3)_4Cl_2]^+$ the Cl-metal-Cl symmetrical stretching vibration produces no change in the dipole moment of the molecule (Fig. 10.2, a and c), and thus the IR spectrum does nc contain any bond corresponding to this vibration.

On the other hand, in the *cis-* isomer of each of these compounds. The symmetrical stretching vibration (Fig. 10.7 b and d) as well as the asymmetrical stretching vibration, produce appreciable changes in the dipole moment. Consequently, a large number of bands due to the Cl-metal-Cl stretching vibration are observed in infrared spectrum of the *cis-* isomer.

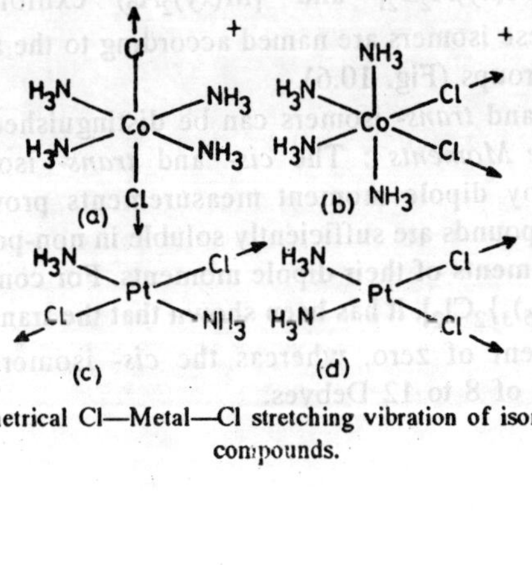

Fig. 10.7. Symmetrical Cl—Metal—Cl stretching vibration of isomeric coordination compounds.

OPTICAL ISOMERISM

Some 6-coordinate complexes can exist as optically active isomers. An optically active substance is capable of rotating the plane of plane-polarized light. Optical isomers are mirror images of each other, but are not superimposable on each other.

Two possible optical isomers of the *cis-* form of the complex $Co(en)_2XY$ are shown in Fig. 10.8. Both of these isomers are capable of rotating the plane of plane-polarized light. One form rotates the plane of light to the right and is said to be *dextrorotatory* and the other form rotates it to the left to an identical extent and is said to be *levorotatory*. The isomers are known as *dextro* and *levo*, respectively. An equimolar mixture of the two isomers becomes optically inactive because of the cancellation of the effect of one isomer by that of the other. This optically inactive mixture is called a *racemic mixture.*

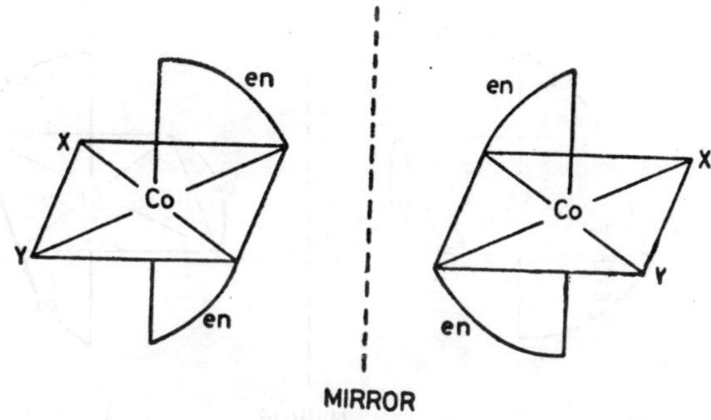

Fig. 10.8. Enantiomeric structures of *cis-* $Co(en)_2XY$.

The term *enantiomers* is used to refer to the dextro and levo forms of a substance. Except for their ability to rotate plane-polarized light differently, enantiomers have almost identical properties.

Optical isomerism among *four coordinate complexes* is encountered only in tetrahedral complexes. Square planar complexes do not exhibit optical activity as they possess a plane of symmetry. Optical isomers of tetrahedral complexes having four different ligands cannot be isolated due to the labile nature of tetrahedral complexes. Tetrahedral complexes of Be, Cu(II), Zn(II), etc. containing unsymmetrical bidentate ligands have been resolved into optically active isomers (see Fig. 10.9).

Fig. 10.9. Optical isomers of bis (benzoylacetonato)beryllium(II).

Some more examples of this type of isomerism in octahedral complexes of the type $[M(aa)_3]$, $[M(aa)_2x_2]$ and $[M(aa)\ x_2y_2]$ (when aa = symmetrical bidentate ligand, x and y = monodentate ligand) are :

(a) $[M(aa)_3]$ type

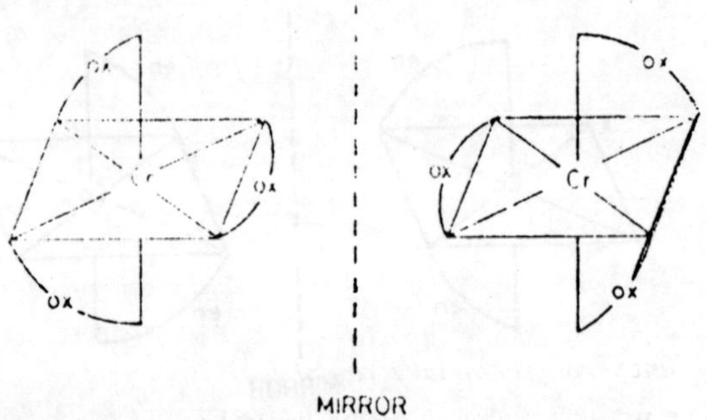

(b) $[M(aa)_2x_2]$ type

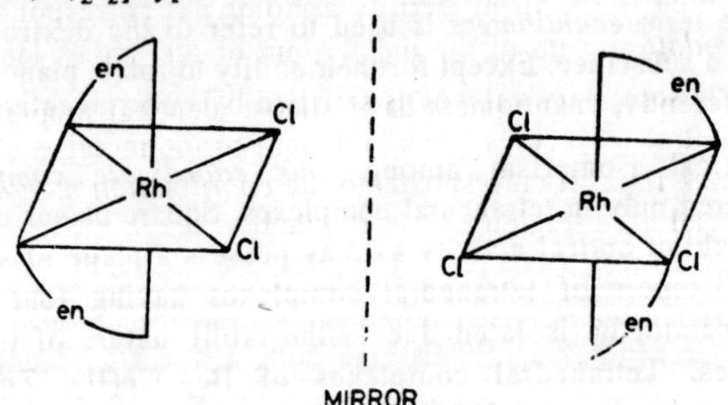

(c) $[M(aa)x_2y_2]$ type

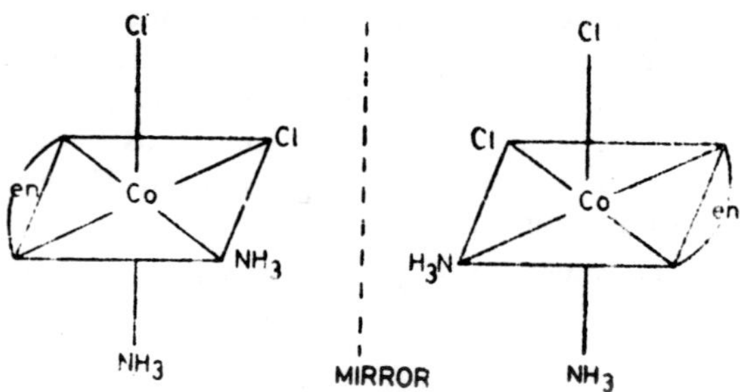

THEORIES OF THE COORDINATE BOND

Werner Theory

The first successful theory explaining the properties of coordination compounds in terms of their structures was proposed by Alfred Werner (1893). Werner's coordination theory, with its concept of primary and secondary valences provides an adequate explanation for the existence of such complexes as $[Co(NH_3)_6]Cl_3$. The properties and stereochemistry of these complexes are also explained by this theory, which remains the real foundation of coordination chemistry.

The salient features of Werner's coordination theory are :

1. Every element exhibits two types of valences, viz., *primary valency* and *secondary valency.*

(i) The primary valency of the central metal ion is satisfied by negative ions. Its attachment to the metal ion is shown by dotted lines. In modern terminology it corresponds to the *oxidation state* of the metal atom or ion and is ionizable.

(ii) The secondary valency is satisfied by either negative ions or neutral molecules. Its attachment to the metal is shown by thick lines. It corresponds to the *coordination number* of the metal atom or ion and is non-ionizable.

2. Every metal has a fixed number of secondary valences. In order to meet this requirement, a negative ion may perform a dual function of satisfying both types of valences.
3. Every metal has a fixed number of secondary valences.

4. The ligands which satisfy the secondary valences are always directed towards fixed positions in space. The geometry of a complex is determined by the number and arrangement of such ligands in space. If a metal ion has a secondary valency of four, the geometry of the complex can be tetrahedral or square planar. On the other hand, if its secondary valency is six, the complex should be octahedral. The primary valences are non-directional.

The complex $CoCl_a \cdot 6NH_3$ may thus, be represented as :

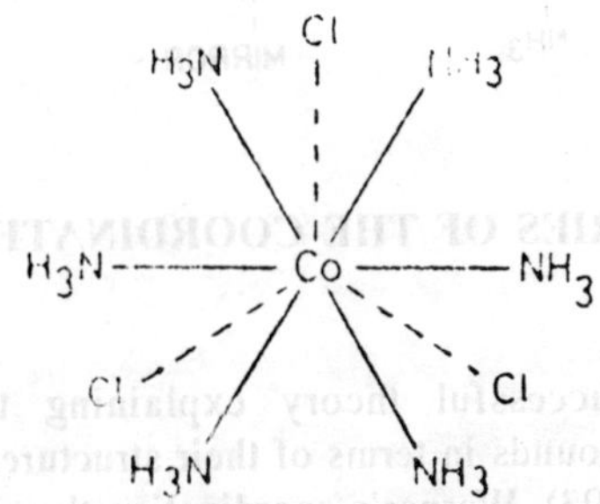

Cobalt has three primary valences and six secondary valences. The three ionizable valences (shown by dotted lines) are satisfied by chlorine and the six secondary valences (shown by dark lines) are attached to ammonia molecules.

In addition to this, three other ammonia complexes of cobalt(III) which Werner prepared, were $CoCl_3 \cdot 5NH_3$; $CoCl_3 \cdot 4NH_3$ and $CoCl_3 \cdot 3NH_3$.

In modern formulation, the complex ion is shown in square brackets with ionizable ligands outside the coordination sphere.

Werner formula	*Modern formula*	Cation	*Anion*	*Total number of ions*
$CoCl_3 \cdot 6NH_3$	$[Co(NH_3)_6]Cl_3$	$[Co(NH_3)_6]^{3+}$	$3Cl^-$	4
$CoCl_3 \cdot 5NH_3$	$[Co(NH_3)_5Cl]Cl_2$	$[Co(NH_3)_5Cl]^{2+}$	$2Cl^-$	3
$CoCl_3 \cdot 4NH_3$	$[Co(NH_3)_4Cl_2]Cl$	$[Co(NH_3)_4]Cl_2]^+$	Cl^-	2
$CoCl_3 \cdot 3NH_3$	$[Co(NH_3)_3Cl_3]$	(non-ionizable)	—	—

Evidence to support Werner's formulation was obtained by (a) precipitating the amount of chlorine present as chloride ion and (b) from conductance measurements.

Werner postulated that the secondary valences are directed towards fixed positions in space. Because of the stereochemical disposition of the ligands (secondary valences), many shapes of complexes are possible which lead to steroisomerism (geometrical and optical isomerism) in such complexes.

For a six coordinated complex, three possible structures are : (i) octahedral, (ii) planar hexagon and (iii) trigonal prism (Fig.10.10). In

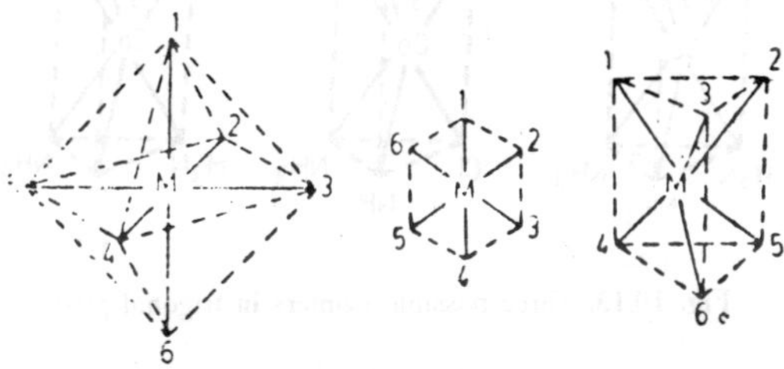

Fig. 10.10. Structures of six coordinated complex — octahedral, planar hexagon and trigonal prism.

these structures, all the six ligands are equidistant from the central metal, M. In the complex $[Co(NH_3)_4Cl_2]Cl$, for example, Werner experimentally isolated two isomers. It means that the experimental result predicts the structure to be octahedral. Only an octahedral structure can form *two* isomers (1,2- and 1,6-) (Fig. 10.11), while planar and trigonal prism arrangements will give three isomers each Figs. 10.12 and 10.13. Similarly, Werner verified the arrangements in other complexes by actually isolating the isomers corresponding to the number theoretically possible.

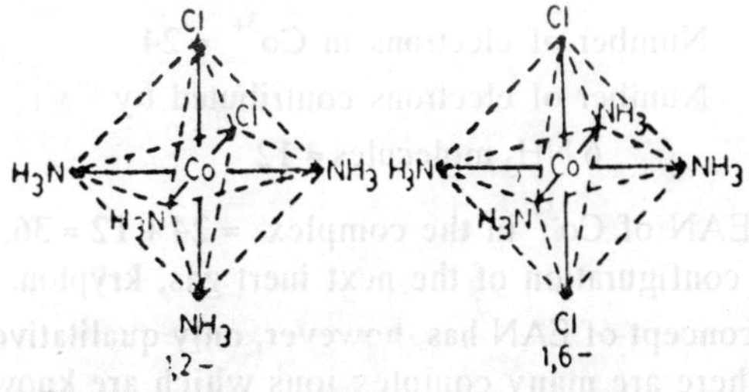

Fig. 10.11. Two possible isomers in octahedral.

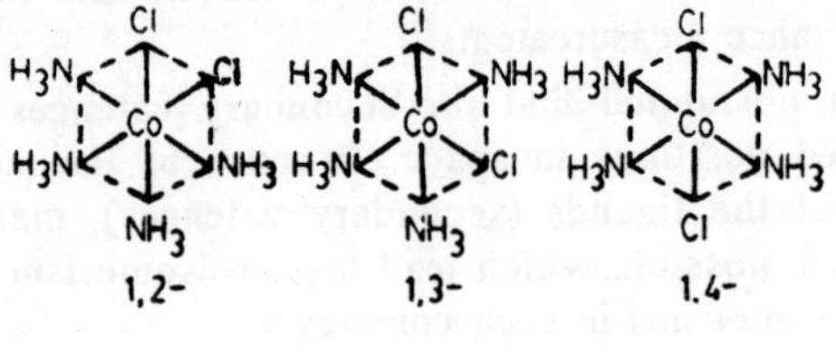

Fig. 10.12. Three possible isomers in planar hexagon.

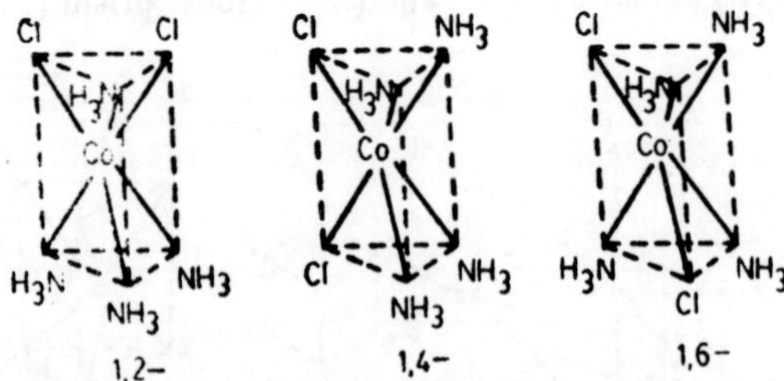

Fig. 10.13. Three possible isomers in trigonal prism.

Effective Atomic Number Rule

Sidgwick extended the Lewis theory to explain the bonding in the coordination compounds. He considered the formation of complexes by donation of an electron from the ligand to the metal ion. The donor atom acts as a Lewis base and the metal ion acts as a Lewis acid. He suggested that the metal ion will continue to accept electron pairs till it achieves the next inert gas configuration. This is known as the effective atomic number rule. The total number of electrons is called the effective atomic number (EAN) of the metal. This can be illustrated by the complex $[Co(NH_3)_6]^{3+}$.

Atomic number of cobalt = 27

Number of electrons in Co^{3+} = 24

Number of electrons contributed by
6 NH_3 molecules = 12

The EAN of Co^{3+} in the complex = 24 + 12 = 36, which is the electronic configuration of the next inert gas, krypton.

The concept of EAN has, however, only qualitative significance because, there are many complex ions which are known to be very stable but do not obey the effective atomic number rule. For example, whereas hexacyanoferrate $Fe(CN)_6^{4-}$ follows the EAN rule, $Fe(CN)_6)^{3-}$ ion does not (Table 10.1).

TABLE 10.1 : EAN OF METALS IN SOME COMPLEXES.

Metal complex	Metal atom	Atomic number	Electrons lost by metal	Electrons gained by complex formation	EAN
$[Fe(CN)_6]^{4-}$	Fe	26	2	6 × 2 = 12	26 – 2 + 12 = 36
$[Co(NH_3)_6]^{3+}$	Co	27	3	6 × 2 = 12	27 – 3 + 12 = 36
$[Ni(CO)_4]$	Ni	28	0	4 × 2 = 8	28 – 3 + 8 = 36
$[Cu(CN)_4]^{3-}$	Cu	29	1	4 × 2 = 8	29 – 1 + 8 = 36
$[Cr(NH_3)_6]^{3+}$	Cr	24	3	6 × 2 = 12	24 – 3 + 12 = 33
$[Fe(CN)_6]^{3-}$	Fe	26	3	6 × 2 = 12	26 – 3 + 12 = 35
$[Ni(NH_3)_6]^{2+}$	Ni	28	2	6 × 2 = 12	28 – 2 + 12 = 38
$[PdCl_4]^{2-}$	Pd	46	2	4 × 2 = 8	46 – 2 + 8 = 52
$[Pt(NH_3)_4]^{2+}$	Pt	78	2	4 × 2 = 8	78 – 2 + 8 = 84

The Werner's theory describes the structures of coordination compounds and explains the existence of various types of isomers, but the nature of bonding within the coordination sphere remains to be explained. For instance, it is not apparent why there are characteristic coordination numbers or varying geometries. Why are some four-coordinated complexes square planar, while others are tetrahedral ?

Most of the transition metal ion solutions in water have characteristic colours. addition of ligands often causes deepening and sometimes changes of colour. How can the changes in colour be related to the nature of bonding in the coordination sphere ? Often the formation of a complex ion is accompanied by striking changes in the magnetic susceptibility of the central metal ion. For example, the ions $[Fe(H_2O)_6]^{3+}$ and $[FeF_6]^{3-}$ have paramagnetism corresponding to five unpaired electrons ; the complex $[Fe(CN)_6]^{3-}$ has only one unpaired electron. It is also desirable to explain how coordination influences chemical reactivity.

The main theories used to explain the coordinate bond are : *valence-bond theory* which assumes the bond to be covalent ; *crystal-field theory* which assumes the bond to be purely electrostatic and offers successful explanation of such phenomena as colour and reactivity ; *I gand-field theory* which is basically crystal-field theory, but with some allowance for a covalent contribution ; and *molecular-orbital theory* which allows for varying ionic and covalent contributions but cannot be applied rigorouly.

Valence-bond theory (VBT)

The valence-bond method, developed largely by Linus Pauling, allows prediction of the geometries of coordination spheres. The central metal ion, acting as a Lewis acid, accepts lone pairs of electrons from ligands into vacant metal ion orbitals.

The stepwise process of formation of a complex may thus be considered as :

1. The loss of electrons by the metal in accordance with its oxidation number.
2. The hybridization of an appropriate number of metal orbitals for the formation of covalent bonds with the ligand orbitals. This will define the stereochemistry of the complex.
3. The occupation of the hybrid orbitals by the lone pair of electrons from the ligands to form σ -bonds.

This can be illustrated by applying the VB approach to an octahedral chromium(III) complex $[Cr(NH_3)_6]^{3+}$:

Chromium(III) has an outer electronic configuration $3d^3 4s^0$, which may be represented as in Fig. 10.14.

	3d	4s	4p
Chromium atom (*ground state*)	↑ ↑ ↑ ↑ ↑	↑	
Cr^{3+} ion	↑ ↑ ↑		

Fig. 10.14.

Fig. 10.14 shows that two 3*d* orbitals, one 4*s* orbital, and three 4*p* orbitals are available to from six hybrid orbitals which can accept six electron pairs from the ligands (Fig. 10.15 where the crosses represent the pairs of electrons shared with the ligands, NH_3).

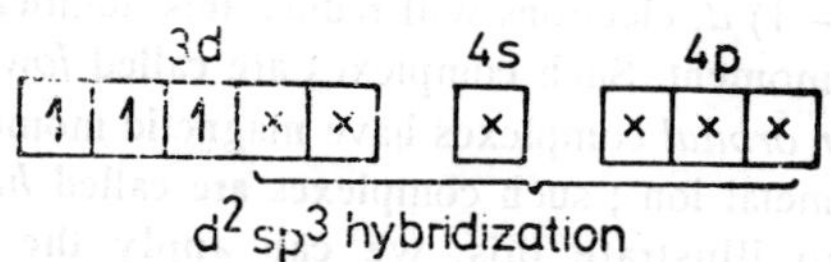

Fig. 10.15. Cr in $Cr(NH_3)_6]^{3+}$ (each of the 6 NH_3 ligands denotes a pair of electrons of Cr).

The complex $[Cr(NH_2)_6]^{3+}$ involves d^2sp^3 hybridization, and its structure is *octahedral* (Fig. 10.16).

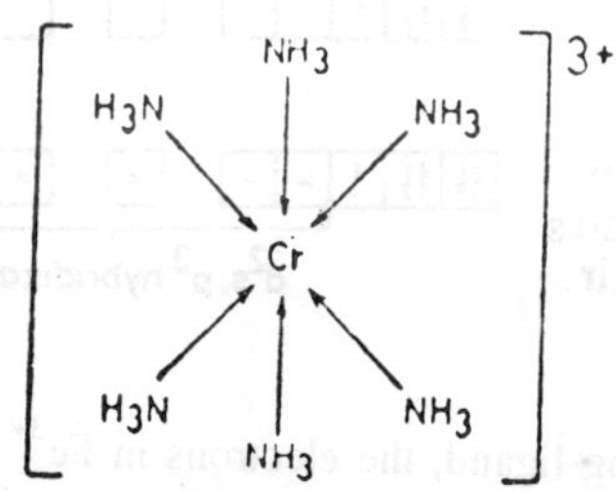

Fig. 10.16. Octahedral structure of $[Cr(NH_3)_6]^{3+}$ complex.

In the case of chromium, the 3*d* electrons present no problem, but as will be seen later, occasions arise when the requirements of bonding conflict with the tendency of non-bonding electrons to occupy the 3*d*-orbitals according to the Hund's rule. In the VB approach, in such cases the non-bonding electrons are forced to pair up against Hund's rule. However, pairing up of electrons takes place only under the influence of a strong ligand. In other cases the bonding electrons tend to occupy 'outer', *nd*, orbitals as against 'inner', (*n* – 1)*d* orbitals in the former case. If the outer, *nd*, orbitals are used, the nonbonding electrons are arranged in the (n – 1)*d* orbitals in the same manner as in the uncomplexed metal ion.

Complexes in which inner, i.e. (n – 1)*d* orbitals are used for bonding are known as *inner orbital* complexes and those using outer (*nd*) orbitals are known as *outer orbital* complexes. The hybrid orbitals are represented as d^2sp^3 and sp^3d^2 respectively. An obvious distinction in the two cases is in the number of unpaired electrons. The use of inner, $(n-1)$ d, electrons will reduce this number and lead to a lower magnetic moment. Such complexes are called *low spin.* One the other hand, *outer orbital* complexes have magnetic moments as high as that of the free metal ion ; such complexes are called *high spin.*

In order to illustrate this, we can apply the valence-bond approach to the two iron(III) complexes, $[Fe(CN)_6]^{3-}$ and $FeF_6^{\ 3-}$:

Applying valence bond theory to $[Fe(CN)_6)^{3-}$:

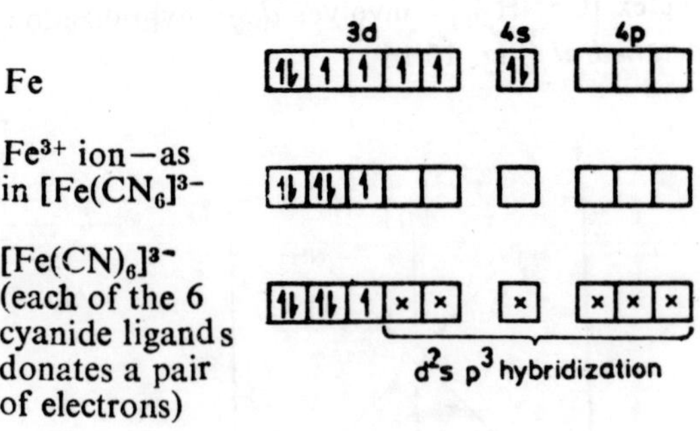

Cyanide is a strong ligand, the electrons in Fe^{3+} ion are paired up against the Hund's rule. In $[Fe(CN_6)]^{3-}$, two 3*d* orbitals, one 4*s* orbital and three 4*p* orbitals hybridize to give d^2sp^3 hybridized orbitals of equal energy. Because of d^2sp^3 hybridization, the structure will be octahedral. There is also one unpaired electron in the complex which will show paramagnetism. Since *inner* 3*d* orbitals are used for bonding, it is *inner orbital complex.* Such inner orbital complexes have low paramagnetism or they are diamagnetic. Thus, it is *low-spin* or *spin-paired* complex.

Similarly, applying the valence-bond approach to $[FeF_6]^{3-}$ complex :

3d 4s 4p

Fe

Fe^{3+} ion—as in $[FeF_6]^{3-}$

$[FeF_6]^{3-}$ (each o the 6 fluoride ligands donates a pair of electrons)

3d 4s 4p 4d

sp^3d^2 hybridization

Fluoride is a weak ligand, the non-bonding electrons in Fe^{3+} ion are placed in accordance with the Hund's rule. In $[FeF_6]^{3-}$ complex one 4*s*, three 4*p* and two *outer* 4*d* orbitals undergo sp^3d^2 hybridization. The structure will be octahedral (sp^3d^2 hybridization). There being five unpaired electrons in the complex, it will be strongly paramagnetic. Since *outer* 4*d* obritals are used for bonding, it will be *outer orbital* (*high-spin*) complex.

TABLE 10.2 : TYPES OF HYBRIDIZATION AND GEOMETRY OF COMPLEXES

Coordination number	Hybridization	Geometry of the complex		Examples
2	*sp*	Linear	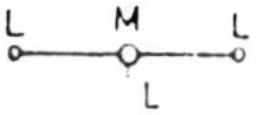 L M L	$[Ag(NH_3)_2]^+$
3	sp^2	Trigonal planar	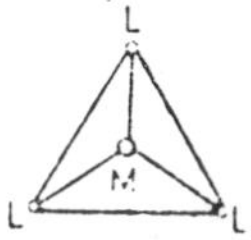 L M L L	$[HgI_3]^-$ $[Cu(PMe_3)_3]^+$, etc.
4	sp^3	Tetrahedral	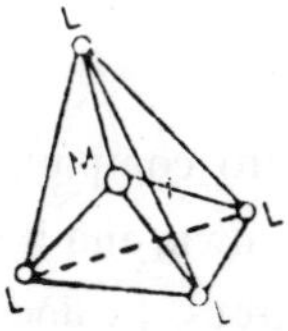 L M L L L	$[VO_4]^{3-}$, $[MnO_4]$, $[NiCl_4]^{2-}$, $[Ni(CO)_4]$
4	dsp^2 or sp^2d	Square planar	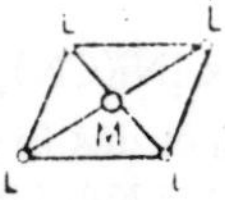 L L M L L	$[Cu(NH_3)_4]^{2+}$, $[Ni(CN)_4]^{2-}$, etc.
5	dsp^2 or sp^3d	Trigonal bipyramidal	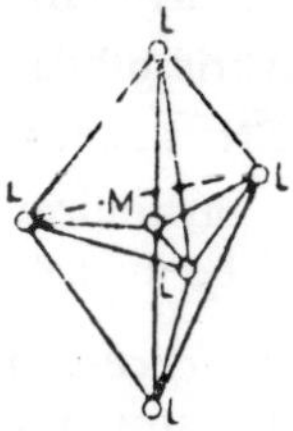 L L M L L L	$[CuCl_5]^{3-}$ $[N(CN)_5]^{3-}$, $Fe(Co)_5$, etc.
6	d^2sp^3 or sp^3d^2	Octahedral	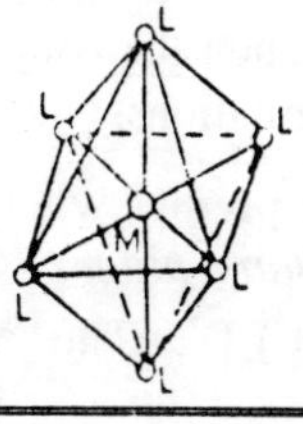 L L L M L L L	$[Fe(CN)_6]^{3-}$, $[Co(NH)_3)_6]^{3+}$, $Cr(NH_3)_6]^{3+}$, etc.

Four-coordinate complexes may have either tetrahedral or square planar structures, depending on the use of different orbitals. The configuration of free nickel(II) and two of its complexes illustrate this point :

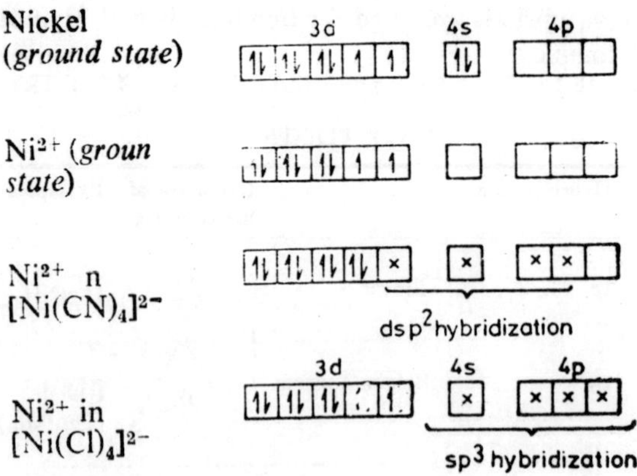

The tetrachloro complex is tetrahedral, corresponding to the sp^3 hybridization shown, and it has a paramagnetism corresponding to two unpaired electrons, as does Ni^{2+}.

In the complex $[Ni(CN)_4]^{2-}$, the ligand is strong enough to force the two unpaired electrons to pair up. The result is that one 3*d* orbital becomes available for hybridization with one 4*s* and two 4*p* orbitals to give four *dsp*^{2} hybrid orbitals. These accept an electron pair from each of the four CN^- ligands. The shape of the complex will be *square planar* (*dsp*^{2} hybridization). There is no unpaired electron in the complex, it will be diamagnetic.

Limitations of valence-bond theory. Although the VBT provides explanation for the structural and magnetic properties of the complexes based upon the concept of hybridization of orbitals, it has certain serious limitations.

1. It cannot predict whether a 4-coordinate complex will be tetrahedral or *square planar.* Consider, for example, a copper(H) complex $[Cu(NH^3)_4]^{2+}$. The VBT predicts the complex to be tetrahedral.

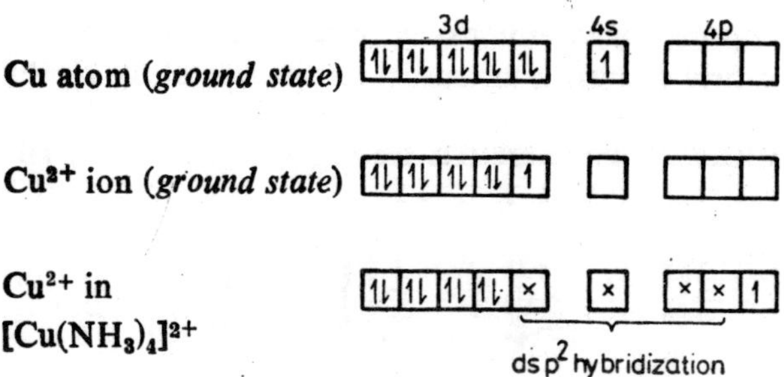

In order to explain, the experimentally determined (X-ray diffraction) *square planar* geometry of copper(II) complexes, it is assumed that one electron from the 3*d* orbital is promoted to the 3*p* orbital to vacate a *d*-orbital for accommodating a pair of electrons from the ligand.

1. One objection to this interpretation is that the odd electron in the 4*p* orbital would be expected to be readily lost giving a copper(III) complex. But this oxidation does not easily occur.
2. For the same metal, in some complexes electrons must be rearranged against the Hund's rule, while in others the electronic configuration is not disturbed. VBT does not offer any explanation for this anomaly.
3. VBT does not predict any distortion in symmetrical complexes, whereas all the copper(II) and titanium(III) complexes are distorted.

In addition to the above shortcomings, the valence-bond approach offers no explanations of the colour observed for complex ions. In a large number of cases, the crystal-field theory goes a considerable way to remedying these defects.

Crystal-field theory

The crystal field theory treats the interaction between the central metal ion and the ligands as a purely electrostatic problem in which the donor atoms act as point charges. Although, it has been shown experimentally that the bonding between ligands and the central metal ion is at least partly covalent, yet the crystal-field theory ignores covalent bonding. The theory was developed by physicists during the

1930's and applied to transition metal ions in a crystal lattice, hence the name crystal-field theory.

Spliting of d-orbitals in octahedral field. To apply the crystal-field approach, the spatial orientations of the *d*-orbitals must be considered, which are shown in Fig. 10.17. In a given free metal ion, all five of these orbitals are degenerate (i.e. they all have the

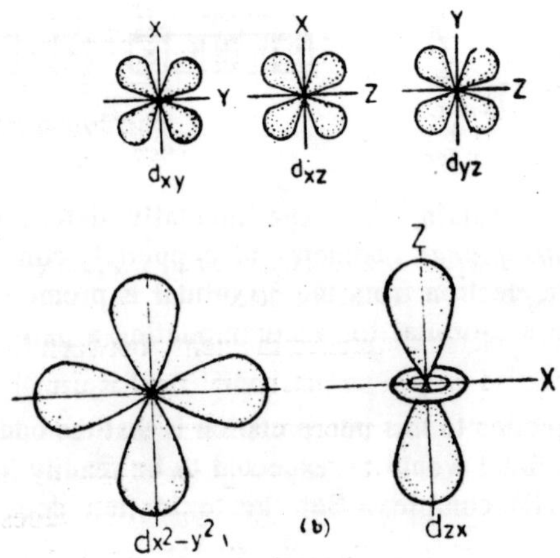

Fig. 10.17. Orientation of *d*-orbitals. (a) t_{2g} orbitals (low energy) ; (b) e_g orbitals (high energy).

same energy.) As the ligands approach the metal ion, the *d*-electrons are repelled by the negative charge (the ligands are considered to carry a negative charge by virtue of their lone-pair) and the energy of the *d*-orbitals increases. If each *d*-orbital were spherically symmetrical, they would all be repelled equally and would remain degenerate as in the isolated atom. But they are not individually spherically symmetrical. The axes of the two orbitals $d_{x^2-y^2}$ and d_{z^2} often designated as the e_g orbitals, coincide with the x^- y^- and z^- axes along which the ligands are approaching. Therefore, when six ligands are brought near the central metal ion along the axes, as in an octahedral complex, electrons in these two orbitals suffer a greater repulsion than the t_{2g} orbitals (d_{xy}, d_{xz} and d_{yz}) which point between the axes. Thus in an octahedral field, the *d*-sub shell is split into a higher energy set of orbitals (e_g) and a lower energy set (t_{2g}). This splitting is shown in Fig. 10.18.

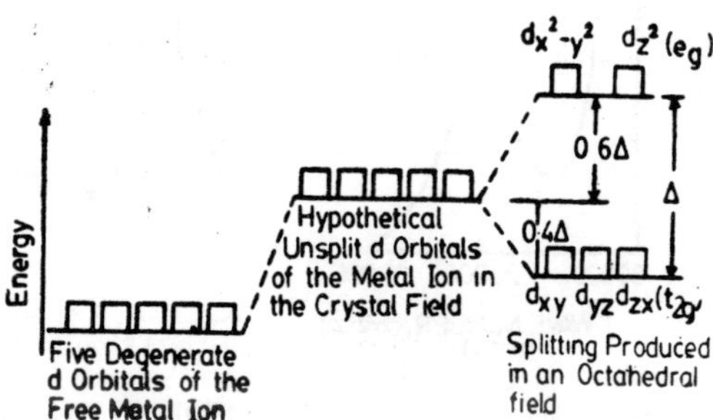

Fig. 10.18. Relative energies of the *d*-orbitals before and after splitting in an octahedral field.

The magnitude of the difference in energy between these two and sets of orbitals (e_g and t_{2g}) in an octahedral field is usually designated as Δ_0. The magnitude of Δ_0 depends upon the metal which is involved, the oxidation state of the metal, the charge or dipole moments of the ligands, and the metal-ligand bond length. In general Δ_0 is more easily determined experimentally than theoretically. The magnitude of splitting is arbitrarily fixed at 10Dq so that $\Delta_0 = 10\,Dq$.

Δ_0 can usually be determined from the absorption spectra of complex ions. In the simplest case (Ti^{3+}ion) when light is absorbed by a complex ion an electron in one of the lower energy, t_{2g}, orbitals is excited to one of higher energy, e_g, orbitals. The energy corresponding to the frequency of the absorbed light is equal to Δ_0.

The absorption spectra of hydrated Ti(III) complex, $[Ti(H_2O)_6]^{3+}$ has a maximum at 20,400 cm^{-1} (Fig. 10.19 a). Thus, for water as the ligand, Δ_0 corresponds to an energy of 244.4 kj mol^1 (1 kJ mol^{-1} = 83.7 cm^+). Since the absorption is in the range of 5000 to 6000 A°, green and yellow components of white light are removed and the observed (transmitted) colour will be a mixture of violet and red. However, Δ_0 cannot always be easily determined from the absorption spectra of complexes. For ions having more that one electron in the *d*-orbitals, there can be other electronic transitions well ; interelectronic repulsions may also exist.

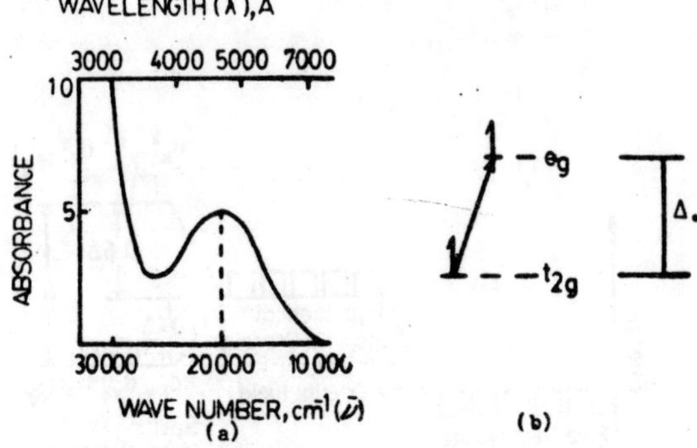

Fig. 10.19. (a) Absorption spectrum of $[Ti(H_2O)_6]^{3+}$ (b) the *d* electron transition in Ti^{3+}

From a measurement of the magnitudes of the crystal field splitting (Δ_0) produced by various ligands with a given metal ion, it is possible to arrange the ligands in a sequence, known as *the spectrochemical series.*

In order of increasing Δ_0 the sequence for some common ligands is as follows :

$$I^- < Br^- < Cl^- < F^- < OH^- < C_2O_4^{2-} < H_2O < NH_3 < en < NO_2^- < CN^- < CO$$

For a given metal ion, complexes containing ligands high in the spectrochemical series will have a large crystal field splitting (Δ_0), than complexes containing ligands low in the series. Thus, for example, for Co(III) complexes with water and cyanide the values are 18,200 cm^{-1} and 33800 cm^{-1} respectively.

Magnetic properties— High and low Spin complex. The electronic configuration of the metal ions, and hence the magnetic properties of transition metal complexes are more rationally explained on the basis of *d*-orbital splitting in the ligand fields.

It is clear from the energy level diagram (Fig. 10.18) that an electron will tend to occupy a lower energy—and hence more stable t_{2g} orbital. Each electron occupying the t_{2g} orbital stabilizes the complex by 0.4 Δ_0 ($-4Dq$) and each electron occupying the higher energy, e_q orbital destabilizes the complex by 0.6 Δ_0 ($-6\,Dq$) units. The gain in energy thus achieved by preferential filling of lower energy, t_{2g} orbitals is called the *crystal field stabilization energy* (CFSE).

The configuration of electrons in the *d*-orbitals shall be ·erned by the following factors in addition to the one mentioned above :

(a) Hund's rule is obeyed because of reduced interelectronic repulsions in different orbitals in a degenerate level.

(b) In a low-spin state, since pairing of electrons takes place, the energy of the system will be raised by *P*, where *P* is pairing energy.

Consider a system with d^1 configuration as in $[TiF_6]^{3-}$. The electron will occupy the more stable t_{2g} orbital and the complex will be stabilized by an amount 0.4 Δ_0 (or $-4\,Dq$). The CFSE for d^2 and d^3 configuration will be $2 \times 0.4 = 0.8\,\Delta_0$ and $3 \times 0.4 = 1.2\,\Delta_0$ respectively. However, for a d^4 ion in an octahedral field, there are two possibilities : (a) either the electron will occupy the higher energy e_g level or (b) it pairs up in the lower t_{2q} level. The CFSE in the first case is $-3 \times 0.4 + P = -1.6 = -0.6\,\Delta_0$, whereas in the second case it is $-4 \times 0.4 + P = -1.6\,\Delta_0 + P$.

Thus, the configuration having four unpaired electrons will be more stable than the other having two unpaired electrons if $\Delta_0 < P$. Hence ligands producing a weak field ($\Delta_0 < P$) will give *high spin* complexes, whereas ligands producing strong fields ($\Delta_0 > P$) will give *low spin* complexes.

Similarly for d^5, d^6 and d^7 configurations both the possibilities of high spin and low spin exist. The possible configuration along with CFSE are given in Table 10.3. The d^8, d^9 and d^{10} configuration, of necessity will have only one possibility irrespective of the ligand field strength (Table 10.3).

Four-coordinate complexes. Crystal field splitting of *d*-orbitals in tetrahedral and square planar fields are shown in Fig. 10.20. In a tetrahedral field four negative ligands lie between the three axes *x*, *y* and *z*. Thus, the three t_{2g} orbitals will point towards the direction of the ligands. Consequently these orbitals will be raised in energy. Since the two e_g orbitals have their lobes pointing between the ligands, they will be lowered in energy. This is in the *reverse order* to what has been observed in an octahedral field. For a given metal ion and a given ligand the crystal field splitting Δ_t for a tetrahedral field is about one half of Δ_0 (octahedral field). In fact, it can be

TABLE 10.3. ELECTRONIC DISTRIBUTION AND CPSE IN STRONG AND WEAK OCTAHEDRAL LIGAND FIELD.

	Strong field				Weak field			
No. of d-electrons in the metal ion	t_{2g} orbitals	e_g orbitals	No. of unpaired electrons	CFSE	t_{2g} orbitals	e_g orbital	No. of unpaired electrons	CFSE
1			1	$-0.4\ \Delta_0$			1	$-0.4\ \Delta_0$
2			2	$-0.8\ \Delta_0$		,2	$-0.8\ \Delta_0$	
3			3	$-1.2\ \Delta_0$			3	$-1.2\ \Delta_0$
4			2	$-1.6\ \Delta_0 + P$			4	$-0.6\ \Delta_0$
5			1	$-2.0\ \Delta_0 + 2P$				$0.0\ \Delta_0$
6			0	$-2.4\ \Delta_0 + 2P$			4	$-0.4\ \Delta_0$
7			1	$-1.8\ \Delta_0 + P$			3	$-0.8\ \Delta_0$
8			2	$1.2\ \Delta_0$			2	$-1.2\ \Delta_0$
9			1	$-0.6\ \Delta_0$			1	$-1.6\ \Delta_0$
10			0	$0.0\ \Delta_0$			0	$-0.0\ \Delta_0$

calculated that for the same metal ion and ligands $\Delta_t = 4/9\, \Delta_O$. The formation of octahedral complexes is thus favoured over that of tetrahedral complexes.

The square planar field can be considered to be one derived from the octahedral field by removing two *trans-* ligands located along the *z-* axis. If the two ligands are not completely removed but are at larger distances as compared to the other four ligands in the *xy* plane, a tetragonal structure results. In such a situation the e_g orbitals will no longer be degenerate. Electrons in the d_z^2 orbital will become stabilized relative to those in the $d_x^2 - y^2$ orbital. The degeneracy of t_{2g} orbitals is also disturbed ; d_{xy} orbital is raised in energy whereas the d_{xz} and d_{yz} are stabilized. The changes are depicted in Fig. 10.20. In a *tetragonal distortion of* the opposite type, in which two *trans-* ligands are brought closer to the metal ion (flattening of an octahedron), the energies of *d-* orbitals will be reverse of this. The sequence of orbitals (increasing order of energy) would then be

$$d_{xy} < d_{xz} = d_{yz} < d_{x^2-y^2} < d_{z^2}$$

Though, the crystal field theory can account for stereochemisty, magnetic properties, colour and other thermody namic properties of

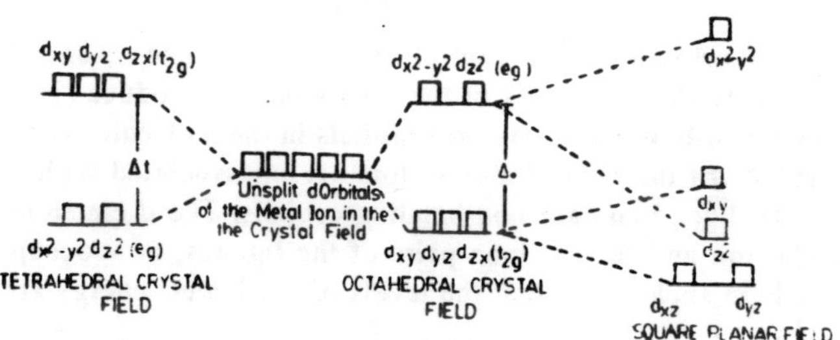

Fig. 10.20. Splitting of *d*-orbitals in various fields.

metal complexes, it only emphasizes the electrostatic nature of metal-ligand bonds. Thus, all properties resulting from covalent bonding of ligands cannot be treated adequately.

Ligand field theory— Molecular orbital approach

Ligand field theory uses the results of crystal field theory but with some allowance for a covalent contribution. Bonding is described in terms of molecular orbitals formed by the interaction of atomic orbitals (liner combination of atomic orbitals LCAO method).

For octahedrally oriented ligands that can form only σ bonds, it is found that six bonding and six antibonding orbitals are formed. In addition, there are also non-bonding orbitals that remain associated with the central atom and play no role in forming the molecule. A qualitative picture for an octahedral complex is given in Fig. 10.21.

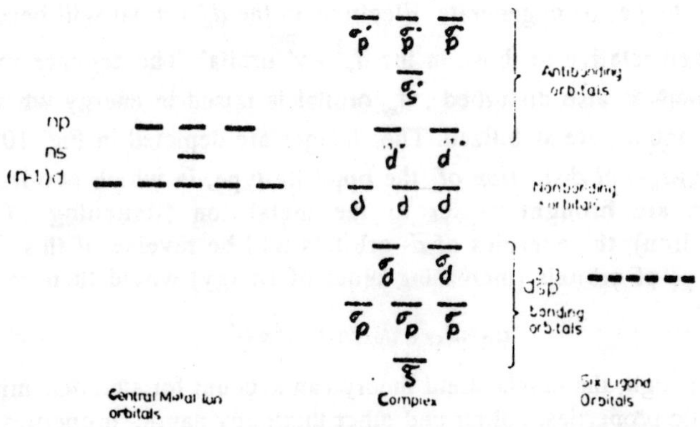

Fig. 10.21. Energies of molecular orbitals in an octahedral complex not involving π bonding.

Six orbitals of the central atom (two *d*, one *s*, and three *p*) combine with six orbitals of the six ligands to produce six bonding and six anti-bonding sigma (σ) orbitals in the molecule. Three of the *d* orbitals of the central metal atom are unassociated with bonding with the ligand and are non-bonding orbitals. The *d* electrons of the central ion and the six lone pairs of the ligands, now occupy these orbitals in such a way that the levels of the lowest energy are filled.

In one respect the ligand field theory confirms the crystal field. That is the separation of *d*-orbitals into three lower levels (t_{2g} non-bonding) and two upper levels (e_g antibonding). The energy separation between the two corresponds to the Δ_0 of crystal field theory. Molecular orbital theory attributes this splitting of *d*-orbitals to covalent bonding.

Pi bonding. Some metal ions and ligands also have the capability of forming pi (π) bonds, π bonding can be of two types.

One is metal-to-ligand π bonding in which the bonding electrons that form the π bond come from the metal. The other type is ligand-to-metal bonding in which the electrons come from the ligand. Metal-to-ligand π bonding occurs in complexes in which the metal ion is in a lower oxidation state, o, I or— I, and the ligand-to-metal π bonding occurs in complexes in which the metal is in a high positive oxidation state. Examples of metal-to-ligand π bonding are provided by carbonyl complexes of transition metals e.g. $Mn_2(CO)_{10,}$ $Ni(CO)_4$ $Mn(CO_5)^-$, etc.

The possibility of π bonding in these complexes arises because CO, in the resonant structure : C : :Ö : has only six electrons on carbon. Hence it can create a vacant orbital that an electron pair from the metal can occupy. Its unshared electron pair, however, makes it capable of forming bonds with vacant orbitals on the metal. Thus, in nickel carbonyl $Ni(CO)_4$, for example, the usual four sigma bonds are formed, giving a sp^3 tetrahedral complex. In addition two π bonds are formed using electrons from the two lower *d*-levels and occupying carbon monoxide orbitals as shown in Fig. 10.22.

Ligand-to-metal π bonding is common in oxyanions such as CrO_4^{2-}, MnO_4^-, etc. All the bonding electrons are imagined to come from the oxide ion.

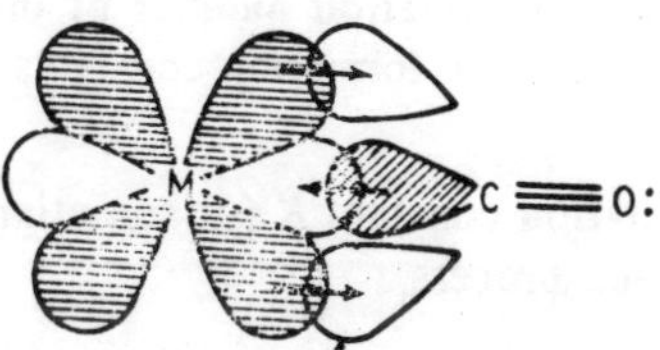

Fig. 10.22. The σ and π bonding orbitals in metal carbonyls (only the M—C bonding orbitals are shown).

STABILITY OF COMPLEXES

Stability has, unfortunately, acquired several loosely defined meanings. In studying the formation of complexes in solution, two kinds of stability have to be considered — *kinetic stability* and *thermodynamic stability*.

Kinetic stability refers to the rate of a particular reaction of complex ions, e.g. substitution, isomerization and racemization. Thermodynamic stability, on the other hand, is a measure of the extent of formation of complex species (under certain conditions) when the system has attained equilibrium. Thus, a complex ion when the system has attained equilibrium. Thus, a complex ion might be thermodynamically unstable with respect to another, but because of kinetic factors the transformation of one into the other could be an exceedingly slow process. On the other hand, a thermodynamically stable complex ion could still exchange ligands rapidly with the surroundings.

Thermodynamic stability. According to J. Bjerrum (1941) the formation of a complex in solution proceeds by the stepwise addition of ligands to a metal. Thus, when a neutral ligand L forms a complex with the metal ion M^{n+}, a number of successive equilibria can be formulated. These may be represented as (charges being omitted) :

$$M + L \rightleftharpoons ML \,;\, K_1 = \frac{[ML]}{[M]\,[L]}$$

$$ML + L \rightleftharpoons ML_2 \,;\, K_2 = \frac{[ML_2]}{[ML][L]}$$

$$ML_2 + L \rightleftharpoons ML_3 \,;\, K_3 = \frac{[ML_3]}{[ML_2][L]}$$

$$ML_{n-1} + L \rightleftharpoons ML_n \,;\, K_n = \frac{[ML_n]}{[ML_{n-1}][L]}$$

where n represents the coordination number of the metal ion, and $K_1, K_2, K_3 \ldots, K_n$ are the stepwise formation constants of the respective complex species.

The overall formation constant K (or sometimes designated as β_n (, corresponds to the process :

$$M(H_2O)_n + NL \rightleftharpoons ML_n + nH_2O$$

and is the product of the stepwise formation constants :

$$K = K_1\, K_2\, K_3, \ldots, nK_n$$

The overall equilibrium constant is related to the total standard free energy change for these steps by the relationship,

$$\Delta G^\circ = -2.303\, RT \log K$$

Since the standard free energy change is related to standard enthalpy ΔH°, and entropy ΔS°, by

$$\Delta G^\circ = \Delta H^\circ - T \Delta S^\circ$$

the thermodynamic quantities can be evaluated by measuring K at different temperatures and using the relation,

$$\Delta S^\circ - \Delta H^\circ\left(\frac{1}{T}\right) = 2.303\,R\log K$$

The magnitude of stability constants for some divalent metal ions of the first transition series with oxygen or nitrogen donor ligands decreases in the order

Mn^{2+} < Fe^{2+} < Co^{2+} < Ni^{2+} < Cu^{2+} < Zn^{2+}

r^{2+} 91 83 82 78 69 74 (pm)

When the same ligand complexes with a metal in two different oxidation states, the one formed with the higher oxidation state is more stable. This is explained on the basis of simple electrostatic consideration. Table 10.4 shows the data for some cobalt complexes in two oxidation states.

TABLE 10.4 : STABILITY CONSTANTS OF SOME COBALT COMPLEXES.

Cobalt(II)		Cobalt (III)	
Complex	*K*	*Complex*	*K*
$[Co(CN)_6]^{4-}$	19.09	$[Co(CN)_6]^{3-}$	64.0
$[Co(NH_3)_6]^{2+}$	4.90	$[Co(NH)_3)_6]^{3+}$	33.66
$[Co(en)_3]^{2+}$	13.82	$[Co(en)_3]^{3+}$	48.70

Chelating ligands such as ethylenediamine form more stable complexes than non-chelating ligands. This can be readily explained in terms of thermodynamic factors. Large values of K (or β_n) are favoured by large negative enthalpy and positive entropy changes. Consider the formation of a complex species containing monodentate ligands in aqueous solution,

$$[Co(H_2O)_6]^{2+} + 6NH_3 = [Co(HN_3)_6]^{2+} + 6H_2O$$

An equivalent number of water molecules are replaced by the ligands. As there is no net increase in the number of molecules in the system, only a small entropy change is expected during the process. However, in a process involving chelating ligands,

$$[Co(H_2O)_6]^{2+} + 3en = [Co(en)_3]^{2+} + 6H_2O$$

each ligand displaces two water molecules leading to a net increase in the number of molecules in the system. Consequently, a considerably large change in entropy occurs. The favourable entropies of complexation are particularly apparent in reaction of EDTA, which has six possible coordination sites—at the two nitrogens and for acetates.

Kinetic stability. In the kinetic sense, it is more appropriate to refer to complexes as being inert or labile rather than stable of unstable.

Inert complexes are those for which reactions occur relatively slowly so that the rate of reaction can be studied by conventional methods. Labile complexes are those which undergo substitution reaction much more rapidly. However, thermodynamic and kinetic stability must not be confused with each other. Thermodynamic stability implies innertness, but the inert complexes need not necessarily be thermodynamically stable. For instance, the complex ion $[Co(NH_3)_6]^{3+}$ is thermodynamically unstable with respect to $[Co(NH_3)_5H_2O]^{3+}$, yet the former persists in aqueous solution for a long time because the substitution of NN_3 by H_2O is an extremely slow process.

APPLICATIONS OF COMPLEXES

Complexes are involved in many practical processes ranging from metallurgical operations to biological process. Some examples are described below.

Cyanide complexes are used in refining of silver and gold from their ores. In the presence of cyanide ions, these metals can be oxidized to their monopositive oxidation states. The soluble complexes are easily separated and the metals isolated. Similarly, the Mond's process for the purification of nickel makes use of the complexation of nickel forming the nickel carbonyl. Pure nickel is obtained by decomposing the carbonyl around 50° C.

There are numerous applications of complex formation in *analytical chemistry.*

APPLICATIONS IN QUALITATIVE ANALYSIS

1. The separation of the chlorides of Ag^+ and Hg_2^{2+} in the first group of the scheme of analysis is based upon the formation of a soluble complex by AgCl with aqueous ammonia.

$$AgCl + 2NH_3 = [Ag(NH_3)_2]^+ + Cl$$

2. Cu^{2+} and Cd^{2+} can be detected in presence of each other by utilizing the comparative stability of their cyanide complexes, namely $K_3[Cu(CN)_4]$ and $K_2[Cd(CN)_4]$. Copper complex is more stable than the cadmium complex so that on passing H_2S into the solution of Cu(II) and Cd(II) ions, in presence of KCN, only Cd(II) gets precipitated as CdS.
3. Fe(III) is detected by the formation of a blood-red coloured complex $[Fe(SCN)]^{2+}$ on the addition of KSCN :

$$Fe^{3+} + SCN^- \rightarrow [Fe(SCN)]^{2+}$$

4. Some other complex formations which serve as confirmatory tests for specific ions are :

$$Ni^{2+} + 2DMG \rightarrow \underset{\text{(red precipitate)}}{[Ni(DMG)_2]} + 2H^+$$

$$Co^{2+} + KNO_2 + HC_2H_3O_2 \rightarrow \underset{\text{(yellow ppt.)}}{K_3[Co(NO_2)_6]^+} + NO + C_2H_3O_2^- + H_2O$$

$$Cu^{2+} + NH_3 \rightarrow \underset{\text{(deep–blue coloured soluble complex)}}{[Cu(NH_3)_4]^{2+}}$$

$$Cu^{2+} + K_4Fe(CN)_6 \rightarrow \underset{\text{(red precipitate)}}{Cu_2[Fe(CN)_6]} + 2K^+$$

APPLICATIONS IN QUANTITATIVE ANALYSIS

Gravimetric estimations of numerous metal ions are carried out by using complexing reagents. For example, the estimation of Ni^{2+} is done by precipitating the ion as $Ni(DMG)_2$ complex, 8-Hydroxyquinoline, also known as oxine, is employed for quantitative determination of a number of metal ions like Zn^{2+}, Al^{3+}, Mg^{2+}, etc. For more examples refer to a text on quantitative analysis.

Some chelating agents, especially EDTA, have proved very useful as reagents for the volumetric estimation of metal ions. A solution of the disodium salt of EDTA when added to a solution of the metal ion, converts the hydrated metal ion into the EDTA complex. Usually the metal ions combine with EDTA in the ratio of 1 : 1. The change in metal ion concentration and hence the end point can be followed by employing a suitable indicator. The indicator is an organic dye which forms a weaker complex with the metal than does the EDTA. As the solution containing the metal ion and indicator is titrated against EDTA, the dye is gradually displaced. At the end point the colour change is from the metal-indicator complex to that of the un-complexed indicator ion.

Many biologically important materials are complex compounds. The hemoglobin molecule is a six coordinated iron(II) complex. Chlorophyll is another important coordination compound. Various enzymes present in the human body are metal complexes.

Bibliography

Bailer, J.C., Jr., (Ed), The Chemistry of Coordination Compounds, 1956, Van Nostrand-Rienhold, New York.

Basolo, F. and Johnson, R.C., Coordination Chemistry, The Study of Metal complexes, 1964, Benjamin, New York.

Cotton, F.A. and Wilkinson, G., Advanced Inorganic Chemistry, 4th edn, 1980, Wiley Interscience, New York.

Durrant, P.J. and Durrant, B., Introduction to Advanced Inorganic Chemistry, 2nd edn, 1970, Wiley Interscience, New York.

Greenwood, N.N. and Earnshaw, A., Chemistry of the Elements, 1984, Pergamon Press, Oxford.

Heslop, R.B. and Jones, K., Inorganic Chemistry — A Guide to Advanced Study, 1976, Elsevier Scientific Publishing Co., Amsterdam.

Huheey, H.E., Inorganic Chemistry, Principles of Structure and Reactivity, 1974, Harper & Row, New York.

Jolly, W.L., The Principles of Inorganic Chemistry, 1976, McGraw-Hill, Maidenhead.

Jones, M.M., Elementary Coordination Chemistry, 1964, Prentice-Hall, Englewood Cliffs, New Jersey.

Kettle, S.F.A., Coordination Compounds, 1969, Thames and Nelson, London.

Lagowski, J.J., Modern Inorganic Chemistry, 1973, Marcel Dekker, New York.

Lee, J.D., A New Concise Inorganic Chemistry, 3rd edn, 1977, Van Nostrand-Rienhold, New York.

Orgell, L., An Introduction to Transition Metal Chemistry, 2nd edn, 1966, Wiley Interscience, New York.

Phillips, C.S.G. and Williams, R.J.P., Inorganic Chemistry, Vol. II, Metals, 1966, Oxford.

Remy, H., Treatise on Inorganic Chemistry, Vols. I and II, 1956, American Elsevier, New York.

Sharpe, A.G., Inorganic Chemistry, 1981, Longman, London.

CHAPTER 11

Lanthanides & Actinides

INTRODUCTION

The inner transition elements are those with underlying (n—2) incomplete f-orbitals. Two series of such elements are known, the first of which is called the *Lanthanide series.* It comprises of fourteen elements (atomic numbers 58 to 71) following lanthanum (at no. 57) and are called *lanthanides* or *lanthanous.* These elements have incompletely filled 4f orbitals and their electronic configurations can be expressed by the general formula $4f^{2-14}5s^2 5p^6 5d^{0-1} 6s^2$. The lanthanides are the largest naturally occurring group in the periodic table with properties so similar that their separation and identification becomes a difficult task.

The considerable confusion related with this group of elements is that of terminology. The name "rare earth" was originally used to describle almost any naturally occurring but unfamiliar oxide. The name then (1920) began to be applied to the elements themselves rather than their oxides. It is now common practice to refer to these fourteen elements from $_{58}$Ce to $_{71}$Lu and $_{57}$La (sometimes Sc and Y as well) as the *rare earths.*

The second series is known as the *actinides series* (5f). Like the lanthanides this series also comprises fourteen elements and is placed in the seventh period in the periodic table. The actinides series follows actinium Ac (at. no. 89) and consists of the elements from (at. no. 90) thorium. Th to (at. no. 103) lawrencium, Lr. All the actinides are radioactive. As per definition the two pairs of elements lanthanum, La [Xe] $5d^1, 6s^2$ and lutetium, Lu [Xe] $4f^{14}5d^1 6s^2$; actinium, Ac [Ru] $6d^1 7s^2$ and lawrentium, Lr [Rn] $5f^{14}6d^1 7s^2$, do not belong to the inner transition series as they do not have incompletely filled *f*-orbitals. They are, however, included in their respective series because of closely related properties.

The two series differ markedly in their properties and are best treated separately.

THE LANTHANIDES

ABUNDANCE

The fourteen lanthanides, historically known as the 'rare-earths' are not so rare. Cerium, for instance, is the twenty sixth most abundant element, being half as abundant as chlorine and five times as abundant as lead.

Over a hundred minerals are known to contain the lanthanides, but only two are of commercial importance. They are :

1. *Monazite*–essentially a mixture of phosphates of La, Th and the lanthanons. It is widely distributed and occurs mainly in Southern India, South Africa, Brazil, Australia and Malaysia.

2. *Bastnaesite*–a fluorocarbonate $MFCO_3$ it occurs in California, New Mexico and Sweden.

GENERAL CHARACTERISTICS

The lanthanide family comprises cerium Ce, praseodymium Pr, neodymium Nd, promethium Pm, samarium Sm, europium Eu, gadolinium Gd, terbium Tb, disprosium Dy, holmium Ho, erbium Er, thulium Tm, ytterbium Yb and lutetium Lu.

Electronic Configuration

The electronic configurations of the elements (free atoms) are determined with considerable difficulty because of the complexity of their atomic spectra. The usual configurations are [Xe] $4f^{n+1}5d^0 6s^3$ or [Xe] $4f^n 5d^1 6s^2$. The configurations are listed in Table 4.1. A comparison of the electronic configurations illustrates the preferred stabilities of half-filled and fully-filled sub-levels. Thus gadolinium has a 5d electron above a half-filed $4f$ orbital. Similarly both ytterbium and lutetium exhibit $4f^{14}$ sub-level configurations.

TABLE 11.1 : GROUND STATE CONFIGURATION OF LANTHANIDES

Element	Symbol	Atomic number	Electronic configuration
Lanthanum	La	57	$4f^0 5d^1 6s^2$
Cerium	Ce	58	$4f^2 5d^1 6s^2$ (or $4f^2 5d^0 6s^2$)
Praseodymium	Pr	59	$4f^3 6s^2$
Neodymium	Nd	60	$4f^4 6s^2$
Promethum	Pm	61	$4f^5 6s^2$
Samarium	Sm	62	$4f^6 6s^2$
Europium	Eu	63	$4f^7 6s^2$
Gadolinium	Gd	64	$4f^7 5d^1 6s^2$
Terbium	Tb	65	$4f^9 6s^2$ (or $4f^9 1d^1 6s^2$)
Dysprosium	Dy	66	$4f^{10} 6s^2$
Holmium	Ho	76	$4f^{11} 6s^2$
Erbium	Er	68	$4f^{12} 6s^2$
Thulium	Tm	69	$4f^{13} 6s^2$
Ytterbium	Yb	70	$4f^{14} 6s^2$
Lutetium	Lu	71	$4f^{14} 5d^1 6s^2$

For cerium, the sudden contraction and reduction in energy of the $4f$ orbitals immediately after La is not yet sufficient to avoid the occupancy of the $5d$ orbital. Hence it prefers a $4f^1 5d^1 6s^2$ configuration to the $4f^2 5d^0 6s^2$ configuration. It is, however, interesting to note that this has a marked effect on the chemistry of cerium which has a well-defined +4 oxidation state, the most common being +3 for almost all other lanthanides.

The $4f$ electrons are sufficiently well-shielded by the intervening electron shells from the surrounding of the atom so that they are not usually involved, in chemical bonding. When slightly excited, one of the $4f$ electrons (less often two) is transferred to the $5d$ level. Their properties are thus mainly determined by $5d^1 6s^2$ electrons and are hence similar.

Oxidation States

All the lanthanides, irrespective of their electronic configuration, exhibit an oxidation of +3. This is due to the fact that the ionization and hydration energies are such that the tripositive state is more stable than the di- or tetra-positive states in aqueous solutions.

Oxidation states of +2 and +4 can be obtained. Ce(IV) and Eu(II) are stable in water and though they are respectively strongly oxidizing and strongly reducing, they have well-established aqueous chemistries. The +4 state for Pr and Tb and +2 state for Nd, Sm, Eu, Dy, Tm and Yb also are known in the solid state but are unstable in water.

The +3 oxidation state is prevalent among the lanthanides as a result of the stabilizing effects exerted on different orbitals by increasing ionic charge. As successive electrons are removed from a neutral lanthanide atom, the stabilizing effect on the orbitals is in the order $4f > 5d > 6s$. The orbitals penetrate through the inert core of electrons towards the nucleus in the same order. By the time an ionic charge of +3 has been reached, the preferential stabilization of $4f$ orbitals is such that in all cases the $6s$ and $5d$ electrons have been removed. Also in most cases the electrons remaining in the $4f$ orbitals are so far embedded in the inert core that they are immovable by chemical means. In case of Ce, however, the $4f$ orbitals are at a comparatively high energy and can therefore lose a further electron.

The occurrence of divalent state in Eu(II) and Yb(II) can be explained on the basis of the stabilizing effects of half ($4f^7$) and completely filled ($4f^{14}$) sublevels.

Atomic/ionic Sizes

On descending a group, the atomic and ionic sizes usually increase because of the addition of extra shells of electrons. On moving across a period, the atomic and ionic radii generally decrease due to an increase in nuclear charge and since the extra orbital electrons shield this charge poorly, all the electrons are pulled closer to the nucleus. The shielding effect increases in the orders $f < d < p < s$. In the lanthanides the shielding of one $4f$ electron by another from the increasing nuclear charge is more imperfect than in the case of d-electrons. Thus, there is successive reduction in size of the elements. Though the decrease in size from one element to another is fairly small, the cumulative effect is large (about 20 pm from Ce to Lu). This is known as *lanthanide contraction.* This gives rise to small differences in properties.

Among the M^{3+} ions a nearly regular decrease is noted from La^{3+} (106 pm) to Lu^{3+} (85 pm). The change in M^{3+} sizes is, however, small from one element to the other and their valency is the same so that their chemical properties are very similar. As a consequence of this contraction, the elements which follow in the third transition series have sizes similar to those of the second transition series. Thus the pairs of elements such as Zr and Hf, Nb and Ta and Mo and W have almost identical sizes and are chemically similar. Consequently their chemical separation is a difficult task. Table 11.2 incorporates the atomic/ionic radii of these elements.

TABLE 11.2 : ATOMIC AND IONIC RADII OF THE LANTHANIDES

Element	Atomic number	Atomic radius (pm)	Ionic radii (pm)		
			M^{2+}	M^{3+}	M^{4+}
Sc	21	164	—	68	—
Y	39	180	—	88	—
La	57	188	—	106	—
Ce	58	182	—	103	92
Pr	59	183	—	101	90
Nd	60	182	—	99	—
Pm	61	179	—	98	—
Sm	62	180	111	96	—
Eu	63	204	109	95	—
Gd	64	180	—	94	—
Tb	65	178	—	92	—
Dy	66	177	—	91	—
Ho	67	177	—	89	—
Er	68	176	—	88	—
Tm	69	175	—	87	—
Yb	70	194	—	86	—
Lu	71	173	—	85	—

Colour

Many of the lanthanide ions have characteristic colours which are not influenced by the anion or by any complexing ligands. The colours due to *f-f* electronic transitions, and are virtually independent of the environment of the ions. Characteristic colours of the M^{3+} ions are listed in Table 11.3.

TABLE 11.3 : COLOURS OF M^{+3} LANTHANIDE IONS.

Lanthanide ion	Number of 4*f* electrons	Number of unpaired electrons	Colour
La^{3+}	0	0	colourless
Ce^{3+}	1	1	colourless
Pr^{3+}	2	2	green
Nd^{3+}	3	3	blue-violet
Pm^{3+}	4	4	rose
Sm^{3+}	5	5	pale yellow
Eu^{3+}	6	6	colourless (absorbs in UV)
Gd^{3+}	7	7	colourless (absorbs in UV)
TB^{3+}	8	6	colourless (absorbs in UV)
Dy^{3+}	9	5	pale yellow
Ho^{3+}	10	4	yellow
Er^{3+}	11	3	pink
Tm^{3+}	12	2	green
Yb^{3+}	13	1	colourless
Lu^{3+}	14	0	colourless

Isoelectronic ions (which have different charges) do not have the same colours. For instance, Sm^{2+} and Eu^{3+} which are iso-electronic are red and colourless respectively. Similarly Yb^{2+} is green, while the isoelectronic Lu^{3+} is colourless. Again Ce^{4+} is orange-red compared to La^{3+} which is colourless.

The absorption spectra of the lanthanide ions contain one or more very sharply defined bands in the UV, visible or near IR regions. The absorption bands of Ce^{3+} and Yb^{3+} ions are broad and apparently result from transitions from the 4*f* to the 5*d* orbitals. Bands for the other ions are line-like in character and arise from transitions within the 4*f* level. The wavelengths of these bands are so sharply and exactly defined that the lanthanide ions are used to prepare optical filters.

Magnetic Properties

The lanthanide ions having incompletely filed 4*f* orbitals would be expected to be paramagnetic due to the presence of unpaired electrons. La^{3+} and Lu^{3+} do not contain any unpaired electrons and thus are not paramagnetic. All other M^{3+} ions are paramagnetic.

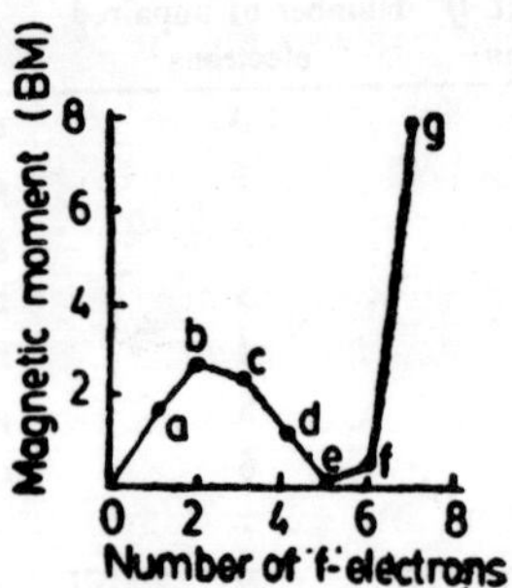

Fig. 11.1

However, the observed paramagnetic moments of the lanthanide(III) ions (Fig. 11.1) do not correspond to the predicted values of the spin only formula. Also, in contrast to the ions of the transition metal ions, the magnetic moments of the lanthanide ions are virtually independent of their environment. This signifies that the 4*f* electrons do not participate in the chemical bonding.

Solubility

Derivatives of lanthanides are generally hydrated. Their solubilities follows the same pattern as the solubility of salts of Group IIA elements.

Their halides (except fluorides), nitrates, acetates and perchlo-rates are generally water soluble whereas fluorides, hydroxides, oxides and salts of oxo-anions, namely CO_3^{2-}, CrO_4^{2-}, PO_4^{3-}, etc. are usually insoluble.

The sulphates of lanthanides are, however, soluble unlike the sulphates of Group IIA metals. Double salts of the type $M_2ISO_4 \cdot Ln_2(SO_4)_3 \cdot 8H_2O$ where M = group IA cation or NH_4 Ln = lanthanide ion, are known.

CHEMICAL REACTIVITY

The lanthanides are strongly electropositive reactive metals. The metals tarnish in air, and if ignited in air or oxygen, burn readily to form Ln_2O_3. Cerium, however, yields CeO_2. The elements also burn in halogens on heating producing LnX_3 and in hydrogen yielding LnH_2 and LnH_3. They react with most one-metals on heating. The metals react with water yielding hydrated oxides. They dissolve in dilute acids rapidly even in the cold forming aqueous solutions of Ln(III) salts.

The chemistry of the lanthanides is dominated by the +3 oxidation state. Because of the large size of Ln^{3+} ions, the bonding in their compounds is largely ionic. Oxidation states of +2 and +4 are also known Ce(IV) and Eu(II) are stable in water ; they are respectively strongly oxidizing and strongly reducing.

COMPLEXATION

The lanthanide ions Ln^{3+} are quite large ions due to which they form mainly ionic compounds. The cations display a strong preference for O-donor ligands. Their coordination chemistry is quite different from, and less extensive than, that of the *d*-transition metals. The coordination numbers are usually high. The stereochemistries, being determined largely by the requirements of the ligands and lacking the directional constrains of covalency, are frequently

TABLE 11.4 : SOME COMPLEXES OF LANTHANIDES.

Oxidation state	Coordination number	Geometry	Example
2	6	Octahedral	LnA (Ln = Sm, Eu, Yb; A = S, Se, Te)
2	8	Cubic	LnF_2 (Ln = Sm, Eu, Yb)
3	4	Tetrahedral	$[Lu(2, 6\ dimethylphenyl)_4]^-$
	6	Octahedral	$[LnX_6]^{2-}$ (X = Cl, Br)
	7	Capped Trigonal prismatic	$[Dy(dpm)_2(H_2O)]$
	8	Square antiprism	$[Eu(acac)_3(phen)]$
	9	Tricapped trigonal prismatic	$[Ln(H_2O)_9]^{3+}$
	12	Icosahedral	$[Ce(NO_3)_6]^{3-}$
4	6	Octahedral	$[CeCl_4]^{2-}$
4	8	Cubic	LnO_2(Ln = Ce, Pt, Tb)
		Square anti-prismatic	$[Ce(acac)_4]$
	12	Icosahedral	$[Ce(No_3)_6]^{2-}$

dpm = dipivaloylmethane $CH_3 \cdot C—\underset{\underset{O}{\|}}{C}—CH = \underset{\underset{O^-}{|}}{C}—C \cdot CH_3$

ill-defined. The complexes are thus, distinctly labile. As a consequence, despite opportunities for isomerism, there appears to be no example of a lanthanide complex displaying any isomerism. Further, complexes with strong usually chelating ligands can be isolated from aqueous solutions. There are generally associated with H_2O molecules. Complexes with unchanged monodentate ligands or with donor atoms other than oxygen are best prepared in the absence of water.

Some typical examples are listed in Table 11.4. Coordination numbers below six are achieved with very bulky ligands. Coordination number six is unusual ; seven, eight and mine are more common. The stereochemistries of high coordination numbers are, in most cases, appreciably distorted.

ISOLATION AND SEPARATION OF LANTHANIDES

Conventional methods are used for concentration of the ores. These are then treated with acid or alkali.

Monazite sand constitutes an important source of lanthanons. It is a mixture of phosphates of lanthanons and also contains thorium phosphate. Monazite sand is digested with sulphuric acid to a paste. It is then leached with water and treated with previously prepared La_2O_3 in the order to reduce the acidity. Insoluble phosphates of Ti, Zr and Th are then separated. The filtrate consisting of solution of sulphates of lanthanous is concentrated with Na_2SO_4 to crystallize the double sulphates of lighter lanthanons. Heavy lanthanons, remaining in the mother liquor, are treated with alkali when $La(OH)_2$ precipitates. These are ignited to get lanthanon oxides.

Cerium is the easiest to separate. This is feasible because Ce(IV) being less basic than La(III) is more easily hydrolysed and precipitates as a basic salt or as the hydroxide when its aqueous solution is oxidized using $KMnO_4$ or bleaching powder. The individual elements are separated by one of the following techniques :

(a) Fractional crystallization

(b) Solvent extraction

(c) Ion-exchange method

(d) Complexation method.

Of these solvent extraction and ion-exchange methods are the most effective. The other two methods serve only to separate the elements into broad groups of light and heavy metals.

The best ion-exchange technique is the "displacement-chromatography" where two separate columns of cation-exchange resin are employed. The first column is loaded with the Ln(III) mixture and the second column (termed the development column) is loaded with Cu(II) ions and the two columns are coupled together. An aqueous solution of a complexing agent, the eluant (a typical one being triammonium salt of EDTA), is then passed through the columns. During the process (Ln(III) is displaced from the first column, and forms complex with EDTA salt. The solution of Ln(III)-EDTA complex reaches the development column where Cu(II) is displaced forming EDTA complex and Ln(III) is redeposited in a compact band at the top of the column. This is possible because Cu(II) being smaller than Ln(III) forms a more stable complex with EDTA. The concentration of $EDTA(NH_4)_3$ has to be controlled to avoid the precipitation of $Cu_2(EDTA) \cdot 5H_2O$.

After the Ln(III) ions have been deposited on to the resin, they are displaced by NH_4^+ ions, in the eluant.

THE ACTINIDES

The fourteen elements, following actinium Ac (atomic number 89) from thorium (at. no. 90) to lawrentium (at. no. 103) are called the *actinides*. They result from the filling of the 5*f* orbitals and are analogous to the lanthanides. Actinium, like lanthanum, is also included in the ACTINIDES series.

Thorium, protactinium and uranium are naturally occurring while the rest are artificially prepared. Elements following uranium are termed transuranic elements.

Every known isotope of the actinides is radioactive and it is only because the half-lives of $^{232}_{90}Th$, $^{235}_{92}U$ and $^{238}_{92}U$ are of the same order of magnitude as the age of the earth that they are still surviving.

Thorium is widely distributed and its commercially important source is the monazite sand. Uranium is also widely distributed ; its most important source is pitch blende, U_3O_8. The transuranic elements have be prepared artificially.

Electronic Configurations

The spectroscopic, chemical and other data indicate that the $7s$, $6d$ and $5f$ states of the heaviest elements of the periodic system are very similar in energy. It is, therefore, very difficult to determine their electronic configurations. For protactinium, for instance, the ground state can be expressed as [Rn] $5f^2\, 6d^2\, 7s^2$ which is so close to the $6d^3\, 7s^2$, $5f^1\, 6d^2\, 7s^2$, or $5f^3\, 7s^2$ state that it is difficult to determine with certainty as to which of the best describes the ground state of the atom.

However, it is insignificant whether these elements possess any *d* electrons in their ground state configurations or not since they will be lost in the formation of tripositive ions–the most common oxidation state. The electronic configurations of the elements are listed in Table 11.5.

Oxidation States

For the first three elements, Th, Pa and U, the most stable oxidation state is that involving all the valence electrons. It decreases

TABLE 11.5 : ELECTRONIC CONFIGURATIONS OF ACTINIDES.

Element	Symbol	Atomic number	Electronic configuration
Actinium	Ac	89	$6d^1\, 7s^2$
Thorium	Th	90	$6d^2\, 7s^2$
Protactinium	Pa	91	$5f^2\, 6d^1 7s^2$ or $5f^1\, 6d^2\, 7s^2$
Uranium	U	92	$5f^3\, 6d^1\, 7s^2$
Neptunium	Np	93	$5f^5\, 7s^2$
Plutonium	Pu	94	$5f^6\, 7s^2$
Americium	Am	95	$5f^7\, 7s^2$
Curium	Cm	96	$5f^7\, 6d^1 7s^2$
Berkelium	Bk	97	$5f^8\, 6d^1 7s^2$ or $5f^9\, 7s^2$
Californium	Cf	98	$5f^{10}\, 7s^2$
Einsteinium	Es	99	$5f^{11}\, 7s^2$
Fermium	Fm	100	$5f^{12}\, 7s^2$
Mendelevium	Md	101	$5f^{13}\, 7s^2$
Nobelium	No	102	$5f^{14}\, 7s^2$
Lawrencium	Lr	103	$5f^{14}\, 6d^1\, 7s^2$

progressively until in the second half of the series, +3 sate becomes dominant. The various oxidation states are listed below :

Ac Th Pa U Np Pu Am Cm Bk Cf Es Fm Md No Lr

+2 (+2) (+2) (+2) (+2) +2

+3 (+3) (+3)+3 +3 +3 +3 +3 +3 +3 +3 +3 +3 +3 +3

+4 +4 +4 +4 +4+4 (+4) (+4) +4

+5 +5 +5+5+5

+6+6+6+6

+7+7

A bold type indicates the most stable oxidation state whereas the unstable states are included in parentheses.

Thorium attains the highest oxidation state of +4 which is also the only one occurring in solution. For Pa, +5 is the most stable. For heavier elements the +3 oxidation state becomes progressively more stable.

GENERAL PROPERTIES

The elements are silvery white metals of high density and relatively high melting and boiling points. The metallic radii decrease sharply from Ac to U (Table 11.6) probably due to an increasing number of electrons being involved in metallic bonding. Thereafter (except Np and Pu) increasing metallic radius is presumably a result of fewer electrons being involved in metallic bonding.

On the other hand, the ionic radii in a given oxidation state fall steadily and indicate that some sort of "ACTINIDE contraction" exists.

TABLE 11.6 : METALLIC AND IONIC RADII OF ACTINIDES.

Element	Th	Pa	U	Np	Pu	Am	Cm	Bk	Cf	Es
Metallic radius (pm)	179	163	156	155	169	173	174	170	185	186
Ionic radius M^{3+}	108	105	103	101	100	99	98	96	95	—
(pm) M^{4+}	99	96	93	92	90	89	88	83	82	—

The metals are reactive and tarnish rapidly in air. They react with most non-metals especially if heated. The elements do not react with alkalies and are less reactive towards acids. Th, U and Pu are rendered passive by conc. HNO_3. The presence of F^- ions prevents this and is the best method for dissolving the metals.

The magnetic moments of some ions of the actinides are shown in Fig. 11.2. The variations are comparable with the tripositive ions of the lanthanides.

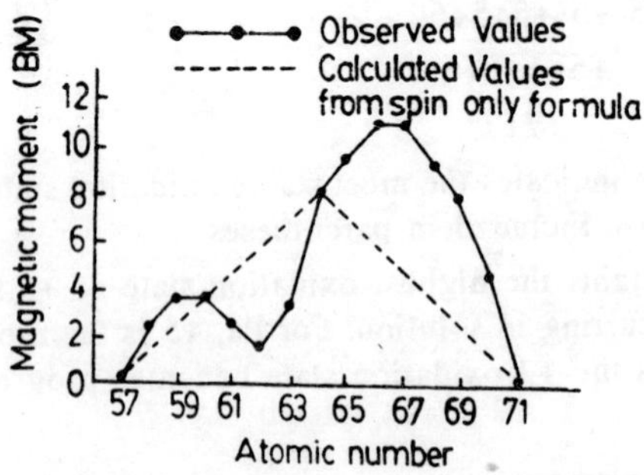

Fig. 11.2. Magnetic moments of some actinides ions.

Thorium is obtained from the mineral monazite sand, while the main source of uranium is pitch blende. Element 93, Np onwards are prepared by typical nuclear reactions.